practical CLASSICS
& CAR RESTORER

Welding Techniques & Welders

KELSEY PUBLISHING LIMITED

Printed in Singapore by Stamford Press Plc Ltd.
on behalf of Kelsey Publishing Ltd,
Kelsey House, High Street, Beckenham,
Kent BR3 1AN (Tel: 081-658 3531).
Under licence from EMAP National
Publications Limited.

ISBN 1 873098 23 5

Acknowledgements

Our thanks to Chris Graham who wrote the
majority of the articles appearing in this book
and to Bryan Dunmall and Mid-Kent College
of Technology with whom he conducted
many of the technique demonstrations and
equipment tests. Thanks also to Michael
Brisby for the gas welding articles, to Paul
Skilleter, John Williams and Gordon Wright
who prepared other articles and to Geoffrey
Ely for technical assistance.

Contents

Introduction

There must have been more coverage given to welding techniques and equipment in *Practical Classics & Car Restorer* than any other motoring magazine – not surprisingly when welding techniques are essential to the restoration coverage that has been the backbone of the magazine.

This book gathers together the pick of the articles that have appeared in *Practical Classics* on the different methods of welding, starting with arc welding which first gave DIY welding a boost in the seventies. Although not the easiest to master arc welding is the cheapest way into welding. The Kel-Arc attachment for arc welders in 1988 made it possible to use arc welding on thinner panels and gave a new lease of life to this system.

By the early eighties gas welding was still considered the only serious method of welding available to the DIY restorer but it was not cheap to buy or to run and it was an elusive art that was hard for the beginner to acquire. The introduction of low priced 'pocket' MIG welders in the mid-eighties brought MIG for the first time within the price range of the DIY enthusiast, and within less than three years brought about a revolution which made it the most popular form of DIY welding. Easier for the beginner to use, more controllable than arc and cheaper than gas, MIG welding proved a boon to the car restorer and is still today pre-eminent in the automotive restoration scene. For this reason more pages of this book have been devoted to MIG techniques and equipment than to any other type of DIY welding.

Not that gas has been ignored. It still remains the favourite of the purist and can provide the finest welding results for those prepared to persevere and master its art.

Practical Classics has also taken a brief look at TIG welding. In the early nineties 'pocket' TIGs became available at prices that the enthusiast could seriously consider. Although it has not yet fulfilled its promise in terms of sales TIG, in combining the ease of use of MIG with the quality results of gas, could be set to do what MIG did in the mid-eighties and bring about another revolution in the hobby scene.

To complete the picture *Practical Classics* also looks at plastic welding and welding with 'glue' (methods that are already gaining acceptance in the motor repair trade). Indeed, this book runs the whole gamut of the viable welding techniques available to the DIY restoration enthusiast, making it probably the most comprehensive work of its kind on the subject.

Arc, Carbon Arc & Spot Welding

Although gas is generally regarded as the most flexible medium with which to weld, many in the DIY fraternity are bound to find it unsuitable. Some of the practical aspects that will undoubtedly influence the budding DIY welder when he comes to equip himself are cost, size, convenience and ease of use, and when these are considered with reference to oxy-acetylene equipment, the picture is not a pretty one.

The alternative however, is to fall back on electricity as a power source — the welding kit in this area being much more portable and readily available. The ubiquitous arc welder is surely known to all these days, and is available in a wide range of specifications and prices. Carbon arc attachments provide the possibility to usefully extend the ability of the arc welder into the brazing field, though you may find them a little harder to track down. Spot welding provides the third electrical option and although being a little more specialised than arc, has its applications for the DIY user.

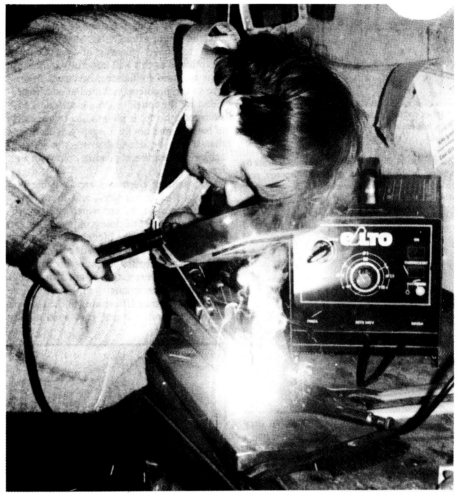

Arc Welding

Arc welders such as the example we chose for this feature, are not particularly complicated pieces of equipment — a fact reflected in that prices start as low as £45. They consist basically of a transformer which can be varied in terms of its output, and which thus determines the 'strength' of the arc being produced. From the welder's casing there appear two leads, one of which is attached to

The controls on the average arc welder are usually pretty basic as in this case.

the hand set, and the other which features a large spring clip; this is the earth or return lead (similar to that on a MIG welder). The latter has to be fastened to the work so that a good electrical contact is achieved; therefore it may first be necessary to clean the area where it is to be clamped.

The hand set is an unassuming looking device with a spring loaded lever that controls the jaws at the tip. Into these jaws is placed the electrode which is consumable, and when the current is flowing, produces the electrical arc between itself and the workpiece. This arc develops a very intense but localised heat which rapidly liquifies the metals to be joined and the electrode itself. The electrodes have a central core of steel which is comparable to the parent metal being welded, and are coated in a flux which forms a protective shield as the molten metal is deposited (again similar to that produced by the inert gas in the MIG technique). General purpose electrodes are commonly available and as their name suggests, they can be used for most jobs. However, another limiting fac-

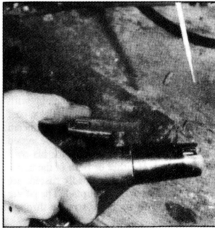

The hand operated lever on the hand set is depressed to open the jaws at the tip, which hold the rod in place.

tor is the current supply which, according to whether it be AC or DC, determines the metal that can be successfully joined.

Arc, Carbon Arc and Spot welding – Chris Graham reports

Arc, Carbon Arc & Spot Welding

The flux coating from the rod is laid down on top of the hot weld and in the ideal case, peels free as it cools . . .

. . . however, if this does not happen, it must be levered or chipped free with the hammer provided – goggles should be worn when attempting this.

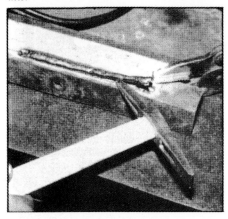

Most of the equipment designed for the home user produce an AC output which is suitable for most ferrous materials, i.e. mild steel, stainless steel, cast iron, etc, using rods which are of course appropriate to the metal to be welded. AC is unsuitable for arc welding aluminium which needs a DC supply. DC sets are considerably more expensive and generally considered to be outside the DIY market.

To start the arc welding process you first set up the joint that you require, by clamping it etc. Arc welding is fairly tolerant with regard to dirt on the joint, but flaky rust should be removed. The sensible technician will usually ensure that the area to be joined is as clean as possible, if only for his own peace of mind. If either or both of the metal subjects to be fastened are badly rusted, arc welding should not be attempted. The power of the arc is likely to 'blow' straight through such samples.

Once the earth clamp has been secured the electrode has to be struck against the subject in order to initiate the arc, and this can be likened to the striking of a match. Once this has been tried by the beginner a few times the skill involved will rapidly become apparent. With the open arc produced, the electrode must be maintained at the correct distance from the work to preserve the arc, and at the correct angle (usually about 60 degrees). This

is made difficult enough anyway by having to peer through the dark green eye shield, but when you remember that the electrode is getting shorter all the time and that it is having to be moved along the joint as well, things become quite tricky. It's always a good idea to have plenty of practice on scrap metal as this will hopefully build both your skill and confidence. An easier alternative to the problems outlined above, is provided by the touch or drag electrode. With such rods the arc is maintained by the angle of the rod, and not by the gap. So this means that it can be dragged along the joint to achieve the same end product, but with a lot less trouble.

Arc welding electrodes are available in a number of differing sizes, and these determine the size of the arc which can be produced, and so the thickness of metal that can be welded. One of the biggest advantages of arc welding is the very limited level of distortion which it creates, because the arc is so localised. However, this is not to say that there are no distortions at all, because there are (as the diagrams illustrate), and distortion can be severe on "thinner" parts such as body panels.

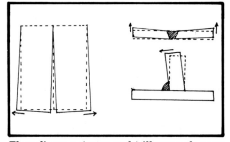

These diagrams (not to scale) illustrate the ways in which different joints distort as they are arc welded.

Generally speaking, arc welding appears to be more suited to work of a heavier nature. It is ideal for use in the fabrication of chassis, axle stands, benches, scaffolding etc. It can be used on car bodywork although the common rod sizes (1.5/1.6mm) take a lot of heat to melt them, which can be too much for the body panel, so to achieve acceptable results calls for practice. The arc welder can be used overhead, downhand (flat on bench) and vertically up or down. Vertical runs obviously create problems as the molten metal tends to

Although thought by many to be rather a basic technique, arc welding can produce some most acceptable results . . .

. . . but care is needed as too much heat can easily 'blow' holes in the subject.

want to drip everywhere. Luckily though there are special rods available for this purpose which help immensely. The other precaution usually taken is that the power is dropped by about 10%, which helps to reduce the 'dribbling' rate.

A further application of arc welding is known as hard facing for which special electrodes are required. This is where a layer of weld is put down to build a worn surface up in preparation for re-grinding as with a crankshaft for example. The teeth on the buckets of JCB diggers are repaired in such a way as this as it is the most economical way of doing so. So do not despair all you owners of classic JCBs, help is close at hand.

Carbon Arc

Once you own a basic arc welder you immediately open the way to another technique known as carbon arc brazing. This process utilises the heat produced from the arc welder via two consumable carbon electrodes, to heat and melt a brazing rod, and make the joint that way. One of the two electrodes is connected to the power, and the other goes to earth, and therefore an open arc is set up between them. The heat supplied

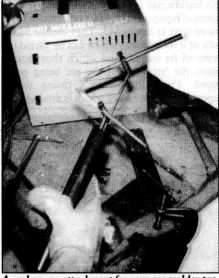

A carbon arc attachment for your arc welder provides a simple and cheap entry into the brazing field. An arc is produced between the two consumable copper coated electrodes which provides the heat for the brazing process – the subject metals are not melted.

should never be sufficient to melt the two metals being joined.

Carbon arc brazing has been described as a clumsy and ungainly technique, and indeed this would seem to be true on first impressions. Dealing with the two consumable rods which come together in a 'V' shape with the arc being produced at the sharp end, is not easy. Older outfits had to be manually operated so that when the rods had burnt too far apart, the brazing had to stop and they had to be re-adjusted by hand to close the gap again. These days however things have been made a little easier because hand sets are now available with controls that enables the rods to be fed in without stopping.

Apart from the initial difficulty of handling the equipment other problems include distortion. This can be quite common with carbon arc work, with one of the contributing factors being that the arc has to be held in one place initially, to warm the metal up and to melt the brazing rod. This gives the heat a chance to spread with sometimes disastrous results. Incidentally, it is possible to cut metal with carbon arc equipment, but this is only with specialised professional equipment that is fitted with a compressed air system to 'blow' through the subject.

Spot welding

Spot or resistance welding relies on a different principle from the common arc in order to make its joint. Electrical resistance provides the heat necessary to fuse the two or more layers of metal together. The spot welder itself is a hand held instrument which is pretty heavy and has two protruding arms. Onto these two arms are fastened the electrodes, which come in many assorted varieties. The body of the spot welder contains a heavy-duty coil together with some other electronic wizardry and sometimes an automatic timer.

The samples of metal are placed between the two electrodes and the machine is activated by depressing the lever on the top. This closes the two electrodes (one on top and the

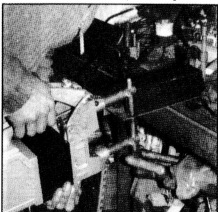

The spot welder, although producing a fine result, is a heavy piece of equipment and so can be a handful to use.

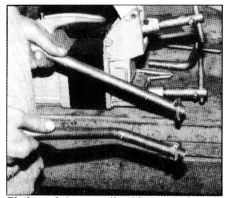

The key to being versatile with a spot welder is to have a large selection of interchangeable electrodes, but these are expensive.

other beneath) onto the metal pinching it tightly together. At the same instant the current begins to flow directly between the electrodes and the resistance which it encounters generates the heat, which melts and fuses the metals together almost instantaneously. The heat produced is very localised and is only applied for a short period, and so distortion is really a thing of the past when you own a spot welder. However, long runs of spot welds can cause problems on thinner subjects so it is best to avoid making such runs by altering your point of attack. Chop and change from one end to the other, the principle involved being akin to tightening a cylinder head.

The electrodes are held tight in the arms by a type of cotter pin fastening. The height of the two electrodes has to be altered in accordance with the thickness of the metal layers being joined.

Preparation is quite important if a lasting weld is to be produced. Basically the relevant areas should be taken back to shiny metal for the best results, and there should certainly be no old paint or rust left to upset the contact. It is also important for the surfaces which are

being joined to be a precise flat fit with each other. Do not simply rely on the pinching action of the welder to achieve this, hand clamps should be used. To be absolutely sure of a tight and slip-free joint, self tapping screws are an excellent idea. They are far less obtrusive than awkward clamps. On the more expensive models a timer is fitted which regulates the duration of the 'power on' stage automatically. This is set in accordance with the thickness of the metal layer under treatment, the relevant information being supplied on a table supplied with the welder. At the cheaper end of the scale I'm afraid it comes down to experience as with no timers fitted on these examples, it has to be judged by the user. Once again practice is a good idea to help gauge the timing, as insufficient heat leads to inadequate fusion and a weak joint, and over enthusiastic heating produces a hole. If holes are created in the workpiece it not only causes problems in that respect, but it also leads to a build up of waste on the electrode. This must not be allowed to occur and to this end special tools are often supplied with the unit for re-shaping the tips.

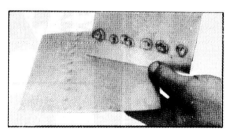

Two examples of spot welding; on the left how it should be done, and, on the right, a beginners attempt. Too much heat is a common failing initially for many newcomers.

Small spot welders are not suitable for use on aluminium but can be used without problems on stainless steel. The use of these welders is limited by the shape of the arms and the electrodes and to be fully versatile you must carry a selection of differently shaped arms which are quite expensive. Problems are also encountered when fixing new panels to old, unless the original metal is spotless. From a purely practical point of view the newcomer to spot welding will soon discover that overhead work is a real struggle. This is not so much due to a difficulty in technique, but simple to the weight of the unit! Everyone adopts their own favourite 'pose' with the spot welder and this is a matter for personal experimentation — if you'll pardon the expression. Finally a word of warning. Electrical welding does have its dangers, especially from shocks, so use your common sense. Avoid welding under damp or wet conditions, and never complete a circuit between the welder, yourself and the earth clamp! □

MIG-ABILITY

NEW SERIES – Part 1

Chris Graham starts this three part series on MIG welding by looking at the principles, preparations and precautions involved.

There is something a little bit frightening about welding. On the face of it, it seems such a violently powerful process that the inexperienced onlooker is bound to think it must be terribly dangerous. In reality, of course, it is not. True, the process does involve molten metal, strange gas and loads of electricity but, assuming you are careful and observe all the rules, you won't get hurt. In most accident cases it is likely that the user will have been at fault and not the equipment. The old saying "familiarity breeds contempt" is worth remembering at every stage during the welding operation because cutting corners, for whatever reason, can lead to injury and wastage and should be avoided at all costs.

MIG welding is, essentially, a very simple process. As with other types of welding the MIG operation relies on the fact that the two samples of metal to be joined are fused together with the help of a filler wire. Gas and arc welding demand that the filler rod be fed in as the weld progresses which means that it is a two-handed process. However, with MIG the wire is introduced automatically from the hand-set, at the touch of the trigger, right to the spot where the weld is being made. It is fed from a reel inside the welder itself, up the cable, to the hand-set where it emerges from the tip. The trigger on the hand-set controls an electric motor which powers the drive rollers to push the wire in a continuous and smooth manner.

Another function of the trigger is to control the gas supply and it is this gas which is the real secret behind the success of MIG welding. Its inclusion in the process ensures a clean and impurity-free weld time after

time. The initial letters which make up the name MIG stand for 'metal inert gas' and it is the word 'inert' that we are interested in here. An inert gas, Argon for example, is one which is unreactive. It will not burn in air or combine with any other elements to form solid compounds.

The presence of such a gas, in the form of a shield around the immediate area being welded, protects the newly laid weld from the formation of oxides and other impurities. If these are allowed to build up then the weld is considerably weakened and, therefore, not as effective.

The other main job of the trigger is to switch on the current. This passes down the wire electrode (made from a similar metal to that being welded) and then arcs from its tip on to the metal surface at the point where the weld is required. As this happens the gas is fed directly to the weld area from the nozzle of the hand-set. Argon is used for welding aluminium but for other types of metal a mixture of gases is essential. Strictly speaking these mixtures, carbon dioxide/argon, argon/oxygen or argon/carbon dioxide, are not completely inert and this leads to the name 'Mag' (metal active gas) being applied to describe the type of welding when they are used. Such situations include the welding of mild steel which requires either carbon dioxide or a carbon dioxide/argon mixture and the welding of stainless steel for which argon/oxygen or argon/carbon dioxide mixtures should be selected.

The electric arc which is struck between the wire and the metal surface being welded is intensely hot. It melts the surface and the wire to form a pool of molten metal which is conducted along the joint as the hand-set progresses. When everything is set as it should be the wire is consumed at exactly the same rate as it is being supplied. The tip is permanently molten and fresh metal from the wire is being 'dripped' into the weld pool all the time. This is difficult to believe until you see it for yourself and, for those of you interested, I would recommend that you watch the video produced by The Welding Institute. This contains some quite superb slow motion close-up photography of the process which proves the point beautifully. Further details are available from the Institute on 0223 891162.

Setting up

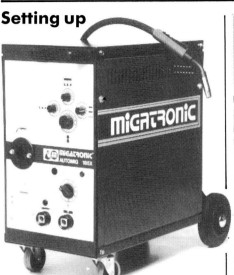

The controls of a typical MIG welder. In most cases they are simple and easy to use.

The first thing that I recommend you do after unwrapping your shiny new welder is to sit down and quietly read through the instruction booklet. This will be invaluable because not only will it explain how to set the unit up for welding but it also should give you an insight into how the process works and to the dangers involved. The benefit of knowing this information will become very apparent the first time something goes wrong. If everything mysteriously grinds to a halt and you have no inkling of what makes the welder tick then what chance is there of you putting it right? On the other hand, if you are well aware of what is involved you will be able to work logically through the possibilities and hopefully sort out the problem.

When you are confident that you understand all the terminology and the procedures involved then you can move on to the next stage. Select the correct gas bottle for the intended type of welding and remove the plastic cap from the top. Check that the threads are clean and then screw on the flow

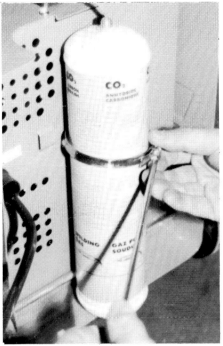

Methods for attaching the gas bottle to the back of these small welders vary but they are always quick and easy.

There is not much to see with the side panel removed. Here the reel of welding wire is about to be mounted on its spindle.

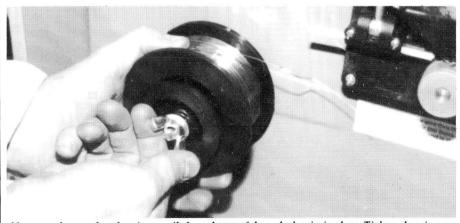

Always make sure that the wire uncoils from the top of the reel when its in place. Tighten the wing nut so it just bites on the reel and prevents it from spinning freely.

control valve. Push in the gas tube firmly and then pull on it to check that it is secure.

The controls on MIG welders are usually very straightforward and I will be explaining them fully in next month's instalment. However, briefly they can be summarised as consisting of an on/off switch and controls for voltage and wire speed. At this stage you should set everything to 'low' and, having plugged in to the mains, depress the trigger on the hand-set to adjust the gas flow. You should refer to your instruction leaflet for details on the setting of this because it is likely to differ from welder to welder.

When fitting the wire reel it is important that it is placed on the spindle the correct way around. The wire should uncoil from the top of the reel so that it can run smoothly into the inlet guide. When tightening up the nut securing the reel you must ensure that there is a small amount of friction. It is very important that the reel is not set in such a way that it can spin freely because this will lead to over-run. When you take your finger off the trigger to finish welding the reel's momentum will prevent it from stopping immediately and unwanted slack will result.

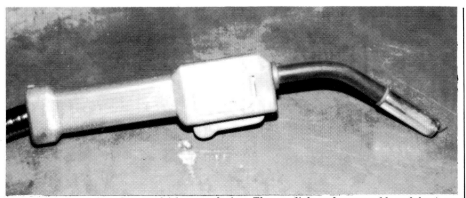

Welding hand-sets these days are fairly smart devices. They are light and manageable and the single trigger controls everything.

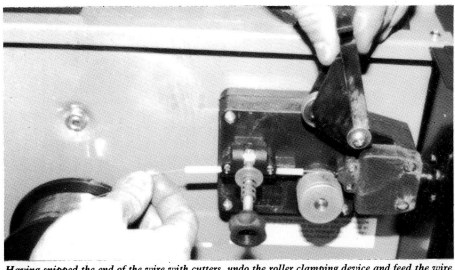

Having snipped the end of the wire with cutters, undo the roller clamping device and feed the wire through the inlet guide tube, across the bottom roller in the groove provided and on into the outside guide tube.

Loosely tighten down the roller clamp and...

Before threading the end of the wire into the inlet guide it should be trimmed with wire cutters to ensure that there are no rough edges to impair its passage. Having passed the tip through the inlet guide, unclamp the drive rollers. Again refer carefully to your instructions at this point as the design differs considerably from model to model. Generally, though, you are likely to find a groove in the bottom roller and it is important that the wire is sitting accurately in this before the top roller is clamped back into position. However, before doing this ensure that the tip of the wire is passed into the outlet tube from which it will enter the cable and proceed up to the hand-set. With this done and the rollers clamped lightly together remove the nozzle and contact tip from the end of the hand-set and, having switched on, depress the trigger until the wire emerges from the top. Trim the end again at this point. Then check that the contact tip is the correct size for the diameter of wire being used – it should be marked accordingly. The final operation is to set the pressure on the wire rollers to the correct degree. The clamp should be wound down until the wire is pushed out of the nozzle smoothly even when a gloved hand is placed in the way.

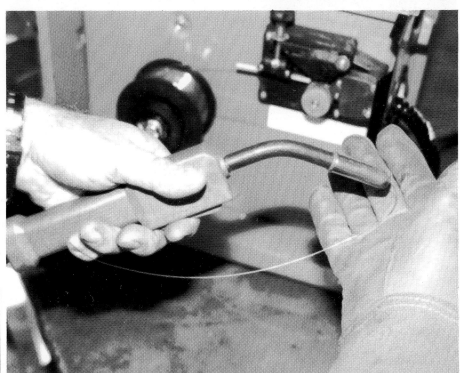

...adjust it until the wire has sufficient force to push its way out against a gloved hand.

Safety

Overalls must be buttoned right up to the neck when welding and synthetic overalls are not recommended — close knit cotton is the best material.

The safety aspects relating to MIG welding are, in the main, derived from commonsense. Never forget that you are dealing with electricity which can both shock and burn you. Ensure that the unit is wired-up correctly and that the correct fuse is fitted within the plug. Avoid obvious mistakes like welding under wet conditions and always make sure that all the side panels are securely fastened to the welder itself before you begin work.

The correct clothing is very important. You should endeavour to cover as much of yourself as possible. The arc produces intense ultra-violet light which will burn exposed skin if it is left unprotected for any length of time. Obviously the overalls you choose should be made from a material which will resist burning from the many sparks produced. Nylon and other synthetic fabrics are useless because they melt and burn. Probably the best idea is close-woven cotton assuming, of course, that it is clean; it is no good using your old overalls which are soaked in oil! Try

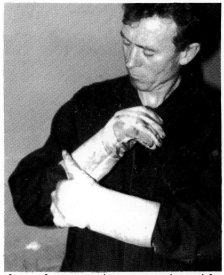

As we always stress in our restoration articles, gloves are vital when welding. Long leather gloves like these are quite suitable.

A hat, again made from a non-flammable material, is also a sensible precaution.

to avoid welding with the overalls open at the neck. If they don't fasten that high then use a clip or clothes peg to make the join.

Hands and wrists are another vulnerable area. Many people I have worked with never even use protective gloves which is asking for trouble. Welding gauntlets are the best answer as they cover the whole forearm to give complete protection. The other important area of the body to be protected is, of course,

If welding in an area where there are likely to be passers-by, erect some kind of screen to protect them from sparks, etc. This is one of the Mid Kent welding areas — notice the heavy curtain.

the head. Never, on any account, weld without a proper face mask. These either can be hand-held or in the form of a hinged helmet. The latter is more convenient because it leaves both hands free for working. The bright UV light from the arc will seriously damage unprotected eyes in seconds so you must make sure that the filter fitted to the mask is dark enough. A numerical scale denotes the strength of a filter and it is wise never to use one for MIG welding which is marked lower than 9. Also, do not forget that anyone else in the workshop or garage must be warned that welding is in progress. Do not let them stand nearby watching in fascination without a protective eyeshield.

Before you start take a look around the immediate area. MIG welding can produce toxic gases which will do harm if breathed in, so always arrange suitable ventilation and never weld in a confined space for any length of time. Look around you to see where the sparks are likely to fall and check for anything flammable. Paint, solvents, petrol cans, gas cylinders, oily rags etc. all present a potential risk. Finally, give some thought to the subject being welded. Disastrous mistakes like welding petrol tanks and close to glass fibre panels have been made in real life so be warned. Also check for coverings of paint, body sealant, filler etc. which could burst into flames if provoked. ☐

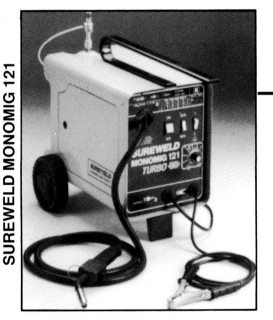

MIG-ABILITY

Part 2: Bryan Dunmall and Chris Graham examine welding techniques and the basic type of joint commonly used.

You may consider that successful welding is all about setting the controls correctly and understanding the processes involved and, of course, this is very much the case. However, there are also some rather more basic requirements which must be met before the 'art' can be mastered. Posture and balance, for example, are both very important. During any welding activity you should be standing comfortably in a well-balanced stance so that your whole body forms a stable base for the welding torch. The use of a full head shield, instead of a hand-held eye protector, is recommended because it leaves you with both hands free. Bryan's chosen technique involves using his spare hand as a support for the one carrying the torch.

The position of your head in relation to the torch is also important. You must be able to see the welding arc and the area immediately around it at all times so that control and accu-

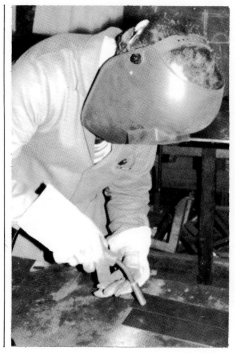

racy can be maintained. If your head is placed behind or above the torch the nozzle will obscure the view. You must position yourself in front or to one side (whichever you find more comfortable) of the torch so that a clear view is ensured.

When using a MIG welder the angle between the subject being welded and the gun (in the direction being travelled) should be, in nearly all situations, about 70 degrees. This relates to the function of the inert gas being supplied through the nozzle. At this angle the weld is afforded the maximum protection which makes it as strong and impurity-free as possible. If the torch is held any shallower the effectiveness of the gas will be reduced and the weld will not be as efficient.

Above: The correct way to do it. Here Bryan demonstrates the right welding position. Note the low position of his head and the use of his left hand as a support for the torch. Gloves, of course, are essential.

Left: This is how not to do it. The head is much too high, meaning that the view of the arc and weld pool will be obstructed by the torch itself.

Another gas-related requirement is that the torch must be maintained at between five and ten millimetres from the surface during welding. This is very important because not only does it aid gas protection but it also controls the length of filler wire protruding from the end of the torch. If this is allowed to be too great the wire is likely to overheat because of its small diameter and the high current being carried. Such a condition causes an unstable arc and a sub-standard weld.

Having struck the arc and begun to weld it must be realised that the speed of travel is a vital consideration. Only with practice will you become aware of the effects of varying this and it is important that you can recognise when you are going wrong. Generally speaking, if you travel too slowly you are likely to burn holes in the metal (especially on thin gauge material). On heavier samples, which will support the slower speed and the excess heat being generated, you will notice a wavy

This picture, kindly supplied, together with the diagrams, by The Welding Institute, illustrates perfectly the way in which the filler wire melts in the arc and drops to the weld pool to build the joint. The height of the torch above the surface is critical and must be maintained accurately for welding success.

effect on the bead as the sign of slow travel. On the other hand, moving too fast will leave a thin bead which lacks penetration. Another useful pointer to watch for is the 'heat

effected zone'. This is the area on either side of the bead which becomes discoloured due to the welding process. With an ideal weld this should stretch for approximately 10mm either side of the bead on samples up to about 1mm thick. On thicker plate it will be slightly narrower and on thinner samples it will be a little wider.

Another point worth noting is that there are two ways that you can weld with the torch. It either can be pushed or it can be dragged. In some books you will find this referred to as leftward and rightward welding but Bryan considers this a little confusing. On very thin gauge metal, where you are trying to avoid burning, 'pushing' is a good idea because the wire is being fed into cool metal all the time as you progress along the joint. On thicker samples, where you require good penetration, 'dragging' is the best method because the wire is then being fed into the hot metal which enhances the welding effect.

Adjustment

MIG welders are relatively simple to control assuming that you understand fully what the adjustments achieve. All sets feature a voltage control and on the cheaper models you are typically provided with a few buttons which can be operated in combination with each other to obtain different settings. In simple terms it is the voltage setting which governs the length of the arc. However, the effects of this are closely inter-related with the other main control which sets the wire speed. The function of this control is self-explanatory but its effects are not quite as simple to assess. Perhaps the best way to understand what's happening is to look at the ways in which alterations in setting effect the weld.

Some manufacturers provide tables giving details of the settings required for varying jobs but it should be stressed that these are intended only as a guide. In practice if the voltage is set too low in relation to the wire speed there will be a short arc and, although the weld will probably penetrate, the bead will look thin and rounded. Alternatively, if you notice that the wire is burning back from the surface and creating a long arc then the voltage is too high compared to the wire speed. The resultant bead will be low and flat and will probably lack penetration. The solutions to these problems rather depend upon the thickness of the material being welded, If, for example, you are working with thin metal sheet and you notice a lot of popping and fizzing the voltage setting you have chosen is too low. However, it would be silly to increase it because you would then run the risk of burning holes. The answer, of course,

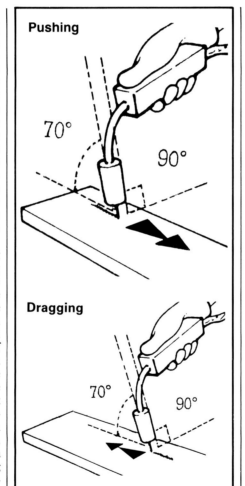

Pushing

70°

90°

Dragging

70°

90°

Pushing and dragging techniques. The choice between these two is dependent upon the job

Checking the bead shape (cross section)

The result of a short arc caused by a high wire speed and/or high voltage

This is the shape of an ideal weld

The result of a long arc caused by a low wire speed and/or high voltage

in this instance would be to turn up the wire speed to achieve the desired balance and the associated even crackle of the properly tuned arc (like bacon frying in the pan!).

Practice is the only way to learn the finer points and Bryan advises that you should always be sure of your settings before attempting the job for real. Very thin gauge plate will require the lowest setting possible but, as the thickness increases, so the voltage can rise accordingly. The method that Bryan adopts is to select the voltage setting he believes to be right for the thickness being welded and then to do the fine tuning with the wire speed control – he says that there is much greater scope for accurate adjustment using this method.

The joints

In this section we are confining ourselves to the basic joints commonly found on vehicle bodywork but, before we start, I think it's wise to say a few words about tacking. Bryan suggests that all joints should be tacked before welding takes place. This will ensure correct alignment and adequate spacing (where required) and make the job far less fiddly. The distance between tacks can be quite large with MIG welding because the heating is so localised and the risk of distor-

tion is consequently greatly reduced. On a 1mm sample of sheet they can be 50-60mm apart. Remember, though, still to be on you guard for distortion and to vary the spacing of the tacks according to the thickness – closer on thinner material and wider apart on thicker samples roughly based on the figures already given. Bryan also considers it advisable to grind away the excess weld from each tack so that they can be welded over smoothly without leaving unsightly bumps.

The two main types of joint encountered on car bodywork are the lap joint and the butt

joint and it is the latter which poses most problems. The butt weld, as its name suggests, involves two pieces being placed edge to edge and welded together. As I have outlined already the torch must be kept at an angle of 70 degrees to the direction of travel but, in addition to this, it must make a 90 degree angle with the surface as seen in the diagram. A butt joint on thin gauge plate is the trickiest of all to undertake. As we discovered in last month's welding supplement, most of the cheaper MIGs are unable to cope with welding thin plate because they cannot

Flat butt joint

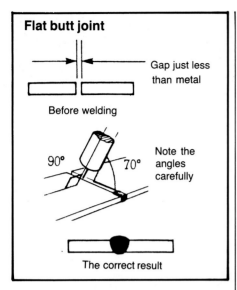

Gap just less than metal

Before welding

90°　70°　Note the angles carefully

The correct result

be set at a sufficiently low voltage level. To overcome this problem your welding technique must be modified. Instead of running straight down the joint in one continuous movement a pulsing action is adopted. You weld for a few millimeters then stop and, as soon as the metal cools to a dull red heat, pull the trigger and continue again, making the set work intermittently down the length of the joint. The idea is that the edges of the metal are allowed to cool sufficiently to avoid the burning effect which destroys the panel.

Flat butt joint (thicker metal)

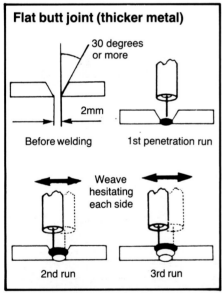

30 degrees or more

2mm

Before welding　1st penetration run

Weave hesitating each side

2nd run　3rd run

Another problem associated with the butt weld on thinner sheet is that the speed required (to prevent burning) is quite high and this can make it difficult to follow the line of the joint. Although, having said that, if you use the pulsing technique this is largely overcome because you are moving more slowly anyway. Many of the expensive sets have this intermittent weld control built-in as a feature. This works automatically and is very useful. On very thin gauge plate the two samples to be joined can be laid tightly together for the weld to be made. However, for 1mm and above you must start to leave a gap in between to aid complete penetration. This is roughly equal to the thickness of the metal being welded. On thicker material the joint has to be 'V'd out' so that penetration is ensured. Using small welding sets for this type of job means that you have to adopt a 'multi-run weld' technique. This involves

laying down a penetration run at the bottom followed by two weaving layers over the top.

If you are butt welding two sections of a panel together and need to have a very accurate joint Bryan knows of a clever way to ensure this. First you should allow the two sections to overlap by about an inch and clamp them tightly together. Then make a saw cut, using a fine blade, down the middle of the overlap to produce the new joint. A useful bi-product of the sawing action is that a penetration gap is created which will enhance the weld and provide a convenient line to follow with the torch.

The lap weld is also quite common and is formed when the two sections to be joined are overlapped and the weld is made into the right-angled corner created. To produce a successful lap weld it is important that the filler wire is aimed directly into the corner which makes the positioning of the torch a critical factor. If it is held too high then most of the weld will be deposited on the top plate

The flat lap joint

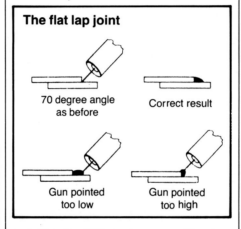

70 degree angle as before　Correct result

Gun pointed too low　Gun pointed too high

and, likewise, if it's too low most will end up on the bottom plate. Either of these two mistakes will result in uneven penetration and a sub-standard weld. The perfect lap weld is produced when the torch is held at 45 degrees so that there is an equal amount of fusion into each side of the joint. It is also important that you try not to melt away the top edge. The direction of travel on lap welds is governed by the thickness of the material in the same way that it is with butt joints. The T-joint is made along similar lines as the diagram shows.

Flat T-joint

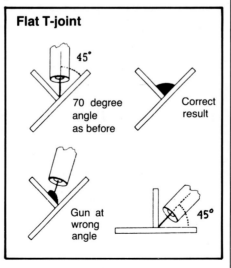

45°

70 degree angle as before　Correct result

Gun at wrong angle　45°

A variation on the butt weld theme is the corner joint. The two sections are welded together at the desired angle with the torch being held in a vertical plane above the joint

Flat T-joint (thicker metal)

For small DIY welding sets 'thicker' means more than 3mm

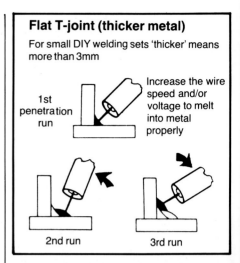

1st penetration run　Increase the wire speed and/or voltage to melt into metal properly

2nd run　3rd run

and at 70 degrees to the angle of travel. Once again a penetration gap is required and it is important that this is right. Make it too wide and the weld will sag through or have it too narrow and penetration will suffer.

Corner joint

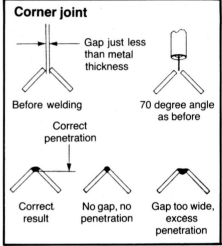

Gap just less than metal thickness

Before welding　70 degree angle as before

Correct penetration

Correct result　No gap, no penetration　Gap too wide, excess penetration

Finally, a useful technique worth mentioning is that of plug welding. It provides a useful substitute for spot welding and can be undertaken with any MIG welding set. Having established where the welds are required a series of holes should be punched on the panel being fitted. After a zinc weld-through primer has been added the panel is set into alignment and clamped firmly. Having cleaned the primer from in and immediately around the hole (with emery paper) hold the torch vertically over the hole so that the filler wire feeds straight down into it. The wire speed should be set fairly high and the voltage turned up little higher than usual. Pull the trigger and make a very small circular action to fill the hole and release it as soon as the weld reached the surface. Plenty of practice on scrap of similar thickness is essential before you tackle the real thing. To begin with it is likely that you will over-fill the hole as you are learning but the excess easily can be sanded away afterwards. The effect of spot welding can be surprisingly well matched using this technique. Professional welding sets are often fitted with attachments to make this job simple and some even produce a slight depression so that the result is identical to a factory-made spot weld.　□

MIG-ABILITY

Part 3: In the final episode of this series Chris Graham investigates some common welding faults and problems.

After reading the instruction booklets and the mounds of advertising literature connected with modern MIG welders, it is easy to lose sight of the fact that good welding is a highly skilled business. Many of the manufacturers would have you believe that all that is required is to open the box, look at the instructions, plug in and start producing perfect welds. Unfortunately, the reality is very different. Beginners are certain to make fre-quent mistakes and encounter many problems before achieving any degree of consistent quality. However, the important thing is to be able to recog-nise these errors and to understand what action must be taken to correct them.

In last month's episode we looked at weld-ing technique and established the importance of setting the welder's controls correctly to suit the job in hand. What follows now will, to a certain extent, be re-covering some of that material to illustrate how poor technique and understanding can lead to welding fai-lure.

The tension on the rollers which provide the drive for the filler wire is adjustable for varying pressures and sizes of wire. It is very important that this is set correctly and that the rollers are not over-tightened. This can lead to problems if the wire snags in the torch, due to some form of blockage, and is prevented from moving further. In such a case the rollers will continue pushing the wire up the cable and it will quickly become buckled and twisted. The result of this will be that the whole length will have to be pulled back out and thrown away which is both wasteful and time consuming. To be sure that you have the rollers set at the ideal pressure, carry out this simple test. Place your gloved hand over the end of the torch and depress the trigger. The motor will operate and, if the tension is set correctly, the rollers will slip on the wire preventing it from becoming crumpled and useless. On the other hand, to ensure that the rollers are not set too loosely, hold your hand about 8in from the torch, pull the trigger and allow the wire to extend out to meet your palm. When it reaches it there should be sufficient power from the rollers for the wire to bend away in an arc and continue feeding. It may take some fiddling to reach this happy medium between the two extremes but it is important that you do so.

The other common cause of this buckling problem is if the return clamp from the earth lead gets knocked off during welding. This will kill the arc immediately but the wire will continue to feed and stub against the surface where it will be stopped dead.

Intermittent wire feed is another cause of trouble. It leads to an unstable arc which makes consistent welding very difficult. This can be traced to a number of causes. There might be dirt in the groove on the feed roller or the groove itself may well have become distorted or badly worn. Alternatively, an obstruction in the lining tube, which guides the filler wire up the cable from the welder to the torch, might be the problem. For example, a build-up of metal powder in the tube can be quite sufficient to produce enough resistance to hinder the free passage of the wire. Oxide build-up on the contact tip can also be a contributory factor. It will effect the movement of the wire and, in severe cases, there is a danger that the arc might jump between the wire and the deposits on the tip to weld the two together, bringing everything to an abrupt halt.

Intermittent wire feed will often be visible to the user who should be able to notice the stop/start action of the wire as the weld is made. In most cases it can be blamed on a lack of maintenance. Infrequent cleaning is the most common cause and the contact nozzle should be regularly changed as a sensible precaution.

Although I have already placed great stress on the importance of correct setting-up this, I am afraid to say, is not the complete answer. Even when everything is perfectly tuned problems can still arise and, generally, these can be attributed to poor technique. Successful welding can be likened to walking on a tightrope. There is a very fine line between right and wrong. The dreaded unstable arc can be produced by something as apparently simple as holding the torch just a few millimetres too far away from the surface. Many people quickly get the hang of welding on a flat bench with test pieces etc., but they find it much harder when welding on a vehicle in anything other than the horizontal plane. This is where experience in being able to recognise the correct torch-to-surface distance is so valuable. The diagrams used last month to illustrate the positioning of the torch in relation to the work surface should be committed to memory. The angles, distances and movements shown must be observed religiously. If the angle of the torch-to-the surface drops away too much then the shielding effect of the gas is lost and the quality of the weld suffers. If the welding torch is held too close to the surface there is a risk that the contact tip will become over-heated, which will ruin it. Remember, you are walking a tight rope!

Probably the first thing that you will all do during the learning process is to burn holes in the metal; everyone does. In nine out of ten cases this will be the result of moving the torch too slowly, particularly when working on thin plate. The speed required for welding thin metal is surprisingly quick, especially when compared to the gas welding technique. Apart from speed, or lack of it, the other main cause of hole-burning is a high voltage setting. Bryan recommends that you should begin with the controls set on low but adds that, in some cases when small MIG units are being used, even 'low' will not be low enough. If you discover this is the case the only option left is to adopt the pulsing technique discussed previously.

The shape of the weld bead produced gives a useful indication into the state of the controls. Being able to 'read' the signs is important. A high voltage, assuming that the metal being welded is of sufficient thickness to resist holes being burnt, produces a flat bead with plenty of associated spatter. A voltage setting which is too low will result in a narrow bead which will lack penetration and can, in some cases, lead to stress cracking.

Generally though, poor penetration is a problem brought about by a lack of welding skill. The most common cause is from inaccurate direction of the torch, although having the wire feed set too slowly in relation to the voltage can also be responsible. On most of the smaller sets you will be using a 0.6mm diameter filler wire (about the size of a needle) and this makes accuracy essential. Failure to aim the torch correctly will produce an ineffective weld. It is very easy to deposit most of the weld on one or the other side of the joint thus making the penetration insufficient.

In the case of butt welds the angle of the torch is important if good fusion and, therefore, a strong joint is to be achieved. It must remain at 90 degrees to both panels so that it acts equally on both sides. The speed of movement of the torch also has an effect on fusion. Travel which is too fast is likely to make a weld lacking complete fusion; however, moving too slowly will either burn holes or lead to an unnecessary and wasteful build-up of weld material. Although this will produce a strong bond you will be throwing away both gas and filler wire and there will be lots of grinding required afterwards. Another disadvantage of slow travel is that a large heat-effected zone will be created. This not only will greatly increase the potential for distortion but it also may subject the surrounding metal to the annealing process which will actually make it softer and therefore weaker.

The other major stumbling block for the novice welder is that of porosity. This is where the weld bead becomes riddled with tiny holes which result from atmospheric contamination (mainly from oxygen). A porous weld is obviously much weaker than a normal one and it is therefore important to realise what is going wrong. On a typical

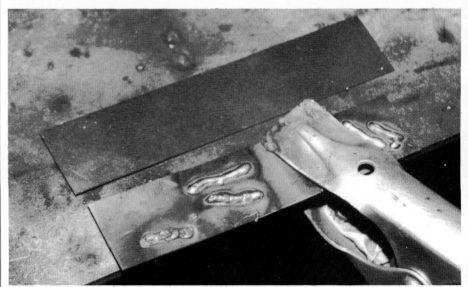

These three short test welds illustrate the welder's performance under different conditions. The bead on the left was made with the gas supply turned off and so is very uneven and porous. The two beads on the right highlight the difference between using an argon/carbon dioxide mix as a shielding gas (top) and pure carbon dioxide (below). The top one is quite obviously the better weld and it illustrates perfectly the improvement which can be had.

occasion when a porous weld is being produced the arc will by much more 'crackly' than usual, there will be lots of spatter thrown out and the arc will be hard to control. Such a condition points to a problem with the gas shielding mechanism and so your first move should be to check the cylinder and its flow meter. If these are all right then turn your attention to the torch. Poorly maintained equipment can suffer from a build-up of spatter in the gas nozzle which can inhibit the flow and lead to porosity. Fortunately, this is not usually a problem to remove and it can be scraped away with a screwdriver or similar. If none of these faults is present then check your surroundings. The MIG process is very sensitive to draughts so welding outside can be awkward on all but the calmest of days.

I would like to conclude with a few words about the choice of gas. I have already mentioned that different gases are used for welding different types of metal (argon for

Summary of welding faults

PROBLEM	APPEARANCE	CAUSE
Poor filling		Torch moving too fast. Current too low in relation to welding speed.
Lack of fusion		Inaccurate use of torch or voltage too low.
Spatter		Voltage too high or a badly cleaned gas nozzle.
Pores		Inadequate gas shielding caused by lack of gas, faulty valve, draughts or a partially blocked gas nozzle.
Uneven joint		Welding too slowly. Current too high in relation to voltage setting.
Poor penetration		Current too low compared to voltage.

aluminium, argon/oxygen for stainless steel and carbon dioxide or argon/carbon dioxide for mild steel) but it is also worth remembering that this choice can effect the amount of spatter produced. This is an important consideration when you are working on bodywork for example. The gas most commonly chosen for general MIG welding is carbon dioxide and the main consideration governing this choice is cost – it is the cheapest. However, it is a relatively active gas which leads to turbulence within the arc. As a result of this the tiny droplets which usually fall from the end of the filler wire into the weld pool get stirred up and many are thrown out in the form of spatter. If, however, a mixed gas (80% argon, 20% carbon dioxide) is selected then this effect is greatly reduced. The argon content gives a much smoother weld and a more stable arc thanks to its inert properties. In addition, another important benefit is that the weld is cooler than with pure carbon dioxide and so the risk of burn through on thin gauge sheet metal is lessened. The one drawback of the mixed gas is that it is more expensive because of the high cost of the argon. □

It all started with him saying that if those smart-Alecs at Practical Classics could do it, so could he.... that was 6 months ago.... now he just sits in here drinking tea and muttering — "Oh God, Oh God"....

Welding Joints

The welded joint is not as simple as you might think — Chris Graham explains, whilst Steve Demol demonstrates.

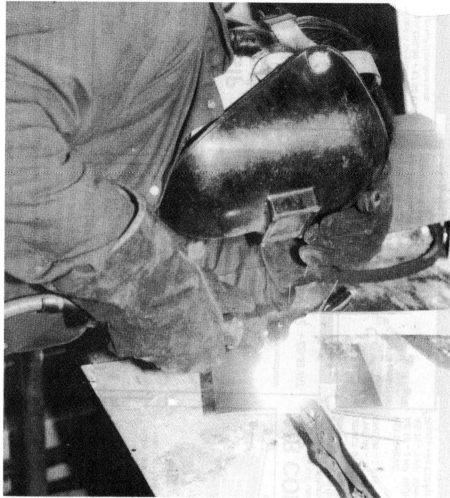

Even when welding small items for illustrative purposes, Steve stresses the importance of taking the correct safety precautions. A full and correctly filtered face shield, thick gloves and suitable cotton overalls should all be worn.

On the face of it you might assume that welding a joint is a very straightforward process, and provided that enough heat is produced to melt the metal, it will be successful. This is not strictly true however; there are very definite methods and precautions which have to be adhered to in order that the weld be correct and long lasting.

Basically there are four major types of welded joint in common use today and these are: the butt weld, the lap weld, the fillet or corner weld and the edge weld. There are of course variations on these joints and we shall touch on some of these later.

Butt joint

The butt weld is achieved, as its name suggests, when two metal samples are butted together end to end, and welded. As with all types of welding, the preparation for this joint is very important. The two edges that are to be joined must run accurately parallel to each other, so that a tight fit is possible. Welding clamps or mole grips should be used to ensure this close proximity. It is no good at all simply to lay the two pieces down side by side on the bench, and try to weld them that way; this is far too inaccurate.

With the clamps in place the next stage is to tack the joint. The frequency of tacks is determined by the length of the joint and the thickness of metal involved – thicker subjects require less frequent tacking. In cases where the tacks come close together, distortion can be a problem and this must be watched carefully. Any corrections should be made with a hammer after the tacking, and before the final welding. It is important to return the tacked joint to as near perfect as possible. If

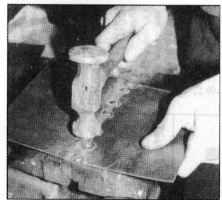

The preliminary tacking of a butt weld can lead to distortion if the subject is thin sheet. Any such distortions should be carefully dressed out with a hammer before the final weld is made.

the subjects are thin sheet then the heat produced by the welding run may be sufficient to 'drag' the two halves above the horizontal and distort the final result. One way around this is to gently bend the tacked joint down to just below the horizontal, so that the welding will then pull it back level. The degree to

which it is bent is obviously critical and can only be learned through experience.

Once the tacks have been positioned, it is a good idea to remove all the clamps to enable you to have a clear run at the joint. It is most inconvenient to have to keep stopping to remove each successive clamp. Another tip worth remembering is that it is often a good idea to place a more substantial piece of metal directly under the joint being welded, when the subject is thin sheet. This will help draw off a lot of the heat and so reduce the effects of distortion.

Lap weld

The lap weld is probably the most common weld to be found on the car, or at least, the basic technique is. Most cars are spot welded together for the sake of speed and efficiency, although the lap weld technique provides the basis for this. One sheet is literally 'lapped'

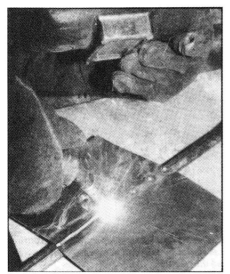

To help reduce the level of distortion when welding a butt joint into a thin subject, it is a good idea to place beneath it a heavier sample of metal. The idea is that the heavier item will take some of the heat straight from the weld, and prevent it from spreading out through the rest of the panel, and causing distortion.

Clamping is vitally important when lap welding. The two edges to be joined must achieve an exact fit together.

over the other (usually by about half an inch), and then they are welded together. Because the spot welder is not a very common piece of equipment in the DIY workshop, a version of lap welding called plug welding can be particularly useful for producing an authentic looking result.

Instead of welding straight down the overlapping joint in the conventional manner, a series of holes is drilled down the edge of the top sheet. Then with the clamps tightly in place, each one is filled with weld. It is best to avoid running straight down the line in succession, but instead to start at one end, then move to the other, then to the middle, and so on. These welds can then be ground smooth and when painted, will take on the appearance of spot welds. Steve says that when filling each hole, he simply runs round the circumference once and this is sufficient to completely fill the hole.

When welding a lap joint in the conventional manner, as in the case of fitting a repair section for example, it is often very helpful to make use of a tool called a joddler. This is basically like a giant pair of pliers which

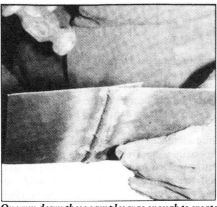

One run down these samples was enough to create this distortion, which illustrates the problem well.

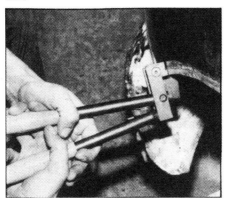

A variation on the lap joint theme involves the use of a joddler. This is particularly useful when fitting a panel repair section as in this case, with the ex-John Williams Jaguar "Mk 1!". The joddler's function is to create a step in the bottom panel so that the repair section can be acurately located, clamped and welded. This type of lap joint leads to a very neat result . . .

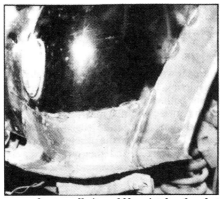

. . . and outwardly it could be mistaken for a butt weld.

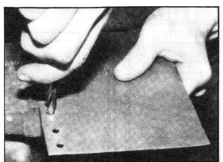

Having drilled the holes in preparation for plug welding, they must be de-burred with a large drill (but keep the gloves on — drills can cut fingers!).

As you might have guessed, clamping is also essential for plug welding. Steve adopts a circular motion to 'fill' each hole . . .

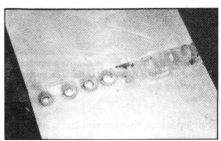

. . . and the results are impressively neat. On the right a few seconds with the grinder has disguised the plugs almost completely.

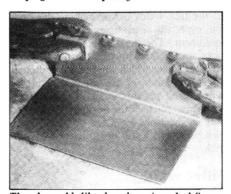

The edge weld, like the others, is tacked first to secure it. Doing this requires a steady hand as the tacks must be put directly on the edge . . .

. . . as must the final weld. When this is cleaned up it produces a very tidy and distortion free joint.

instead of simply squeezing, moulds the metal to produce a small step. This is used on the bottom panel to enable the top one to sit neatly into it. The weld can then be run around the edge of the step once it has been tacked, and then ground off flush with the level of the panel, to produce an 'invisible' join – sounds simple doesn't it! Once the tacking has been completed check carefully

Welding Joints

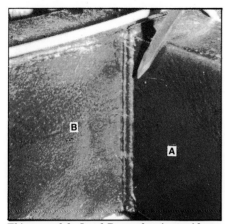

This area of the Land-Rover chassis provides an excellent example of a fillet weld. The outrigger (A) is welded to the main chassis rail (B) where it touches.

sonal preference. He finds it a lot easier with the MIG or the Arc.

With this method tacking is important and distortion can be a problem. When for example, you are fixing a tube to a base, make sure that the tacks are evenly spaced around the circumference. Shrinkage is also a problem, especially when welding single plate. As the newly welded joint cools it can shrink and pull the component towards the weld, so once again, calculated bending at the tacking stage may be advisable. In some cases the situation may arise where a much thinner gauge of metal is being welded onto a thicker sample. Under these circumstances care is needed to avoid distortion in the thinner piece. The one precaution that can be taken against this is to play the heat on to the heavier metal as much as possible.

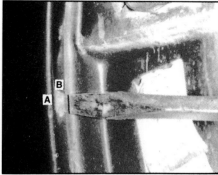

An obvious example of a lap weld on the Spitfire is provided by the joint between the 'B' post (B) and the rear wing (A).

on all lap joints, that the two layers have not separated due to the heat. Any such gaps will be hard to bridge with weld, and will probably result in an untidy finish. Another useful tip to remember is that it is sensible to grind off the tack welds before making the final run, especially when using electric welding equipment. The reason for this is that the tacks are likely to provide that little bit of extra resistance which will prevent complete weld penetration of the metal thicknesss. When using gas this is not such a serious problem as you have the flexibility to play the heat on to the tacks for a few seconds longer each time, which will be sufficient to do the trick.

Fillet weld

The fillet or corner weld occurs when the end of one sheet comes in contact with the side of another, and is welded there. Another example is provided when a tube is fixed to a flat sheet by welding around its base. Fillet joints are encountered on cars in areas such as the chassis and the sills. Steve recommends that such welds be done with electrical equipment rather than with gas, but this is only his per-

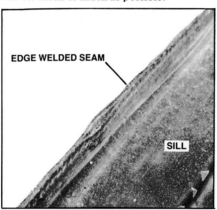

A common example of an edge weld can be found on cars where replacement sills have been fitted, like our Spitfire for example. In this case there are three separate layers of metal that have been edge welded together.

Edge weld

The edge weld is the last type that we shall be considering here. It is achieved when two sheets, both with returns, are butted together so that these returns meet, and they are welded along this inside edge – hence the name. The returns on the panels are normally set at 90 degrees, and they provide a very

good method of joining the sheets. This method is virtually distortion free as the heat is effectively prevented from running through the rest of the panel by the returns. A typical example of this type of weld can be encountered when fitting replacement sills. With edge welding clamping is vital as a perfect fit must be achieved down the whole length. Steve would advise that if you own or have access to gas equipment, use it for this type of welding. Often, he says, you can get away without needing to use any filler wire, as the joint can simply be fused together as it stands. This of course does depend upon the accuracy of the fit between the two returns, so it is well worth spending a lot of time dressing out any imperfections. If everything goes correctly, a very neat join should result and there will be no grinding required.

For all of these types of welding a sensible and methodical approach to the preparation is a very good idea. All the hints and tips concerning this and the important safety aspects which we have pointed out in earlier episodes in this series, should be taken note of before any work begins. □

Welding Alloy Castings

Got a broken or cracked alloy casting? Chris Graham looks into some repair methods

The technique of casting has been around for a long while, many hundreds of years in fact, and over this time the principles involved have remained much the same. Of course there are alternatives to casting as a method of component production but these do not compare very favourably on practical grounds, in today's economic climate. Both forming the item from a solid block of metal or making it up from accurately shaped

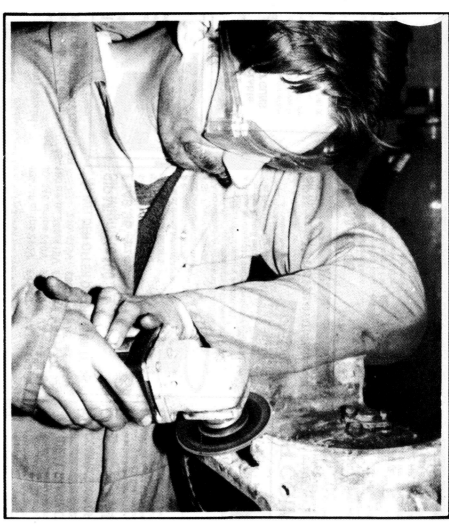

Large castings such as this Austin 7 crankcase with its shattered clutch housing are well within the bounds of practical repair. However, the secret is knowing exactly what it is made of so that the correct welding rods can be chosen – without them the job is virtually impossible.

pieces are ludicrously expensive and time consuming in comparison.

Casting involves producing the desired item from a mould that is made using special sand. A 'pattern' which accurately represents the component (usually made from varnished wood) is sunk into the sand which is then packed tightly around it. The pattern can then be removed to leave a perfect impression which is filled with molten metal. This is allowed to solidify and the casting is made. However, sacrifices are made for the sake of speed and efficiency and the result is that castings are brittle and will not generally withstand shock or impact of any kind, and they are also prone to cracking if exposed to sudden changes in temperature, so damage of one sort or another is fairly common. When this does occur the repair can be made by arc welding, gas welding, MIG or TIG welding or brazing. Iron is often used for casting but in general, it is more troublesome to repair. For a start it can be expensive. If arc welding is your chosen method, a packet of suitable rods is likely to cost about £60. Gas and TIG

Preparing the area properly is important. Any loose or cracked segments should be removed (with a hammer if necessary) and then the edge ground to a fairly smooth and clean finish.

can be used as alternative methods as can brazing (using a silicon/bronze rod), but some iron castings simply cannot be welded. One of the hardest tasks is identifying the type of castings you have. If you are in any doubt about this, it is probably advisable to seek knowledgable opinion elsewhere on the matter. A further problem is caused by the

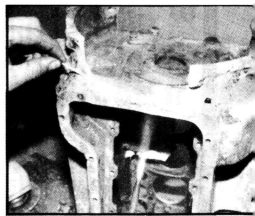

Having prepared the area the next stage is to pre-heat it. In this instance we used a gas welding torch but a household blow lamp should be just as effective. Note how the match is being used to 'test' the temperature. Unfortunately though we did not get much further with this example as Steve found it impossible to lay down the first run onto which the rest of the repair could be built. The casting seemed to contain sand and other impurities which prevented a weld being made.

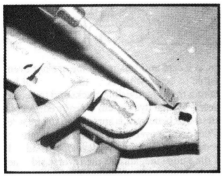

The damage to this cast aluminium water manifold is typical of that found on elderly castings (it too is from an Austin 7). The casting as a whole had taken on a light matt grey appearance which pointed to an oxide build-up – similar to an anodised component. The metal around the hole was wafer thin and it was crumbly, which made it very hard to repair with weld. The only alternative would have been to sleeve it and make it good that way.

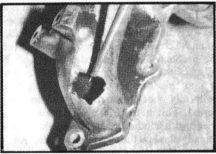

This manifold on the other hand (off a 'modern' Alpine I hasten to add) proved to be much more of a repairable item, and to begin with the edges of the hole were cleaned...

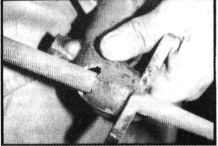

...using a variety of different files...

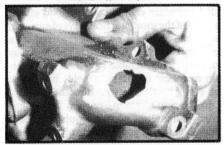

...to produce a finish ready for welding.

fact that the iron has to be heated considerably before welding can begin, and this is usually beyond the ability of the DIY blow lamp.

One of the most commonly used metals in the production of castings is aluminium and this poses its own problems when repairs are required, which are due to the metal's ability to conduct heat so readily. This property causes the localised heat supplied by the welding torch to spread rapidly throughout

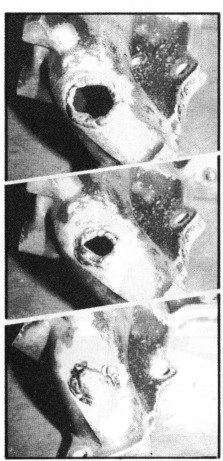

One of the approved methods of making such a repair involves working the weld round and round spiralling inwards towards the centre, as this sequence shows.
The torch can be used to smooth out the finished weld to some degree. Careful application in a similar circular motion can improve the result no end. Remember that the finished weld must not be quenched in any way for fear of cracking the casting.

the whole casting, making actually reaching the correct welding temperature difficult. The answer of course is to pre-heat the area. On the motorcar some typical examples of castings are the cylinder head, the block and the bell housing, and with large items such as these, a lot of pre-heating is required. Smaller articles such as the thermostat housing will obviously require less heating, and in some cases, none at all.

In the case of the damaged aluminium casting, the area should be heated to about 450°C, and there are certain guides that can be used to indicate roughly when this temperature has been reached. One method involves dragging a match (not the striking end) across the heated area and observing whether or not it leaves a black carbon trail. A similar test can be performed with a piece of soap. The crucial thing however, is to avoid over-heating the sample as it is easy to melt it. In most cases the DIY type gas flame gun should be enough, but a gas welding torch may need to be called upon. You should also remember to heat gradually, bearing in mind how easily castings can be cracked.

Obviously the degree of damage to be repaired can vary from a few cracks to a gap-

ing hole, but there are techniques for dealing with most eventualities. The problem is in deciding by which method you are going to effect the repair, though in most cases this decision will be made for you by the equipment that you have available. Aluminium has to be welded using DC equipment when using Arc equipment and such sets are far from common amongst the DIY owners.

The reason for this is a rather technical one, but is basically due to the 'stop/start' nature of the AC current (as opposed to the 'continuous flow' of the DC), which creates a heating and cooling (and cleaning) effect. When using DC equipment, the 'cleaning' action is provided by the flux coated rods, and this is essential to prevent oxidation. As a general rule, magnetic metals (those containing iron) can be welded with AC equipment, but the non-magnetic ones require a DC supply. Gas equipment is a more readily available but slow alternative and does require the addition of aluminium welding flux and suitable rods. The choice of rods is of major importance as the wrong one simply will not weld.

Both MIG and TIG (Tungsten Inert Gas) are particularly useful if large holes are to be filled as they are both fast and do a good job but TIG is the best as it combines great flexibility with speed to produce a very clean result.

However, if the method you have chosen requires the use of welding rods, you must decide positively which type you need, and the first step in this choice is to determine exactly what metal or metal alloy you are dealing with. Each variation will require a welding rod of different metal 'make up' – there is a large selection available on the market. It is a good idea to talk to someone 'in the trade' about what types of rods they use for different jobs, or failing that, to find out from the original manufacturers what, exactly, the casting is made of. Then you can take this information to your local welding suppliers and they should be able to advise you from there.

A pure aluminium casting will usually require an aluminium/silicon rod (with 5% or 10% silicon) to weld it successfully, but if the aluminium has been anodised the rod will probably need to be either pure aluminium or an aluminium/magnesium alloy (5% magnesium). Anodising is a process whereby the aluminium casting is coated in a layer of oxide film (a few microns thick) using an electrolytic process. This greatly increases its resistance to corrosion but has a detrimental effect upon its affinity to the welding torch. In fact such a coating has to be removed before welding can begin, and a grinder is required for this. The coating is very tough and it will require a lot of effort to remove it.

Preparation

Regardless of the technique being used, the preparation before welding begins is very important. As we have discoverd in previous welding articles, the ultimate success of the weld depends on whether or not you have clean metal on which to work. This is obvi-

Welding Alloy Castings

ously not such a problem on aluminium castings as they do not rust, but oxides and dirt can accumulate and need to be removed. Angle grinders, rotary files, hand files and chisels are all useful tools to have handy at this stage, and it is also a good idea to splash on some de-greasing solution as well.

If the repair that you have undertaken involves the 'sticking back together' of a broken component, you may discover further faults along the broken edge during the cleaning stage. Blow holes which are like small air bubbles within the casting can become apparent, and if you do not grind these out, you should at least make sure that they are clean. In some castings (especially older ones) you will find sand within the metal and these will have originated from the sand packing at the actual casting stage. Close inspection will be needed to detect this

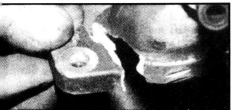

This broken mounting lug provided another fairly typical need for a repair. Some people might imagine that as the two pieces are an exact fit with each other as they stand, that this is how they should be welded back together, but this is not the case.

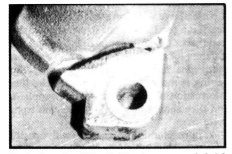

Both edges should be filed back to a smooth finish and slightly 'Veed' (the top of the crack made slightly wider than the bottom. This will help the weld to penetrate to the base of the joint easily, thus ensuring an even weld. In this case setting the two up for welding was easy – a flat surface was quite adequate.

fault but it is worth the effort. If they are left in place during welding they will rise up to the surface as the metal becomes molten, to cause all sorts of problems. They should be ground back to 'clean' metal as soon as they are spotted. The main objective at this stage is to produce a clean and smooth edge but at the same time, retaining its basic shape. *If you use a grinding wheel to clean up an aluminium casting never use that same wheel on steel subjects – it could explode as the aluminium fragments clog the wheel and cause it to overheat.*

In cases where there is a crack to be welded up it should be drilled at each end to halt it

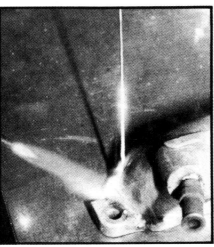

Keep the welding rod well coated in flux at all times.

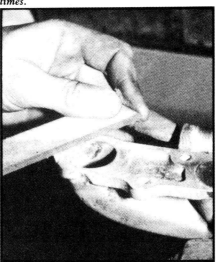

With some careful filing...

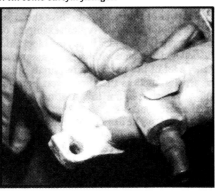

...an invisible joint can be the happy result.

and then gouged out slightly. The holes should be plugged with weld and their size varied according to the casting thickness – thicker casting, larger hole. The gouging is necessary in order to widen the crack out so that its base can be reached by the weld. It's no good for example, simply running the weld along the top of the crack and not allowing it to penetrate to the base, as this will not stop the crack from continuing its downward progress. However, all cracks are not as sim-

ple to locate and repair – some are not even visible to begin with. A casting may well be leaking through invisible cracks and to detect these a special dye system is required. A coloured dye penetrant is sprayed on to one side of the casting over the area that is thought to be damaged. Then a dye developer is added to the same side, and on the reverse side the results can be watched for. With time, the dye should percolate through and become visible thus locating the cracks exactly.

Heat problems and finishing

As I have mentioned already, the casting is not at all happy when subjected to sudden extremes in temperature, and this fact can cause problems at both the welding and the cooling stages. Cracking while you weld is a distinct problem and can be very discouraging – welding one crack well only to find that you have created half a dozen others! A lot of the problem here is caused by the 'on/off' effect of the welding process, but one way around this is to try and keep the sample uniformly hot throughout the welding period. This is best done with the aid of a friend with another gas torch, which he continually plays on the area to reduce the temperature fluctuations.

Cooling the casting after welding is the other problem and often people make very elementary errors. Never quench the casting in any solution or use an air line to cool it. Rapid methods of cooling such as these spell disaster as fresh cracks will inevitably result. Even a cool draught in the workshop can be enough to cause damage. It is essential that cooling is done in a controlled and gradual manner. Gas flames can be used to slow down the cooling rate but this is a time consuming and laborious method. Sometimes heavy metal boxes are used to provide a controlled atmosphere for the casting, but probably not many of us have large, heavy metal boxes in the workshop so this is of little help. A sand filled container can be used to good effect, but one of the most practical ways of ensuring crack-free cooling is by using the oven of an ordinary electric or gas cooker. This idea may encounter considerable resistance from certain members of the household, but if you can swing it, it's a very good method.

In the case of casting repairs, finishing is largely a matter of what is required by each individual job. If you can possibly do so, it is a good idea to leave the repair 'as welded', and not to do any grinding back at all. The reason for this is that there will again be a risk of cracking when using the grinder. However, if for cosmetic reasons the weld must be finished off to a smooth and undetectable finish, then care will be needed. □

Brazing & Soldering

I feel honour bound to confess here and now that so far my only real experience of soldering involved some rather dodgy repairs made to some even dodgier Scalextric cars, many years ago. Casting my mind back even further, to those halcyon days spent in the school metalwork shop, I dimly remember being told about "these 'ere irons". The objects in question had elderly and mis-shapen wooden handles, long metal shafts and then at the tip, a large lump of what appeared to be rusty metal. They were certainly not my idea of an iron especially as the heads were actually made of copper. So it was with great interest that I, together with Steve Demol, launched into this month's feature, the results of which are recounted to you here.

The iron is in fact the solderer's most important tool. It is the device that once heated, is used to transfer the molten solder on to the job, where it sets hard to secure the joint. The head of the iron is made from copper for the following two reasons. Copper is a first class conductor of heat and also allows its surface to become easily 'tinned' (a process that will be explained later). Special soldering stoves are available for heating the irons, but it is possible to achieve satisfactory results using the flame from a domestic gas cooker. It is important to guard

The act of brazing has to be treated with respect in terms of safety. Goggles should be worn (and gloves too — we must remind Steve — but not nylon clothing) and it should only be attempted in a well ventilaged room. Brazing overhead is not a good idea as gravity easily overpowers capillary attraction.

1

To simply join the two sheets of plate together (as in the construction of tubular housing etc), a sweat joint is the ideal solution. First the two plates involved should be coated in solder paint (or tinned in the conventional manner) to cover the touching surfaces.

against overheating as this causes a bit to become pitted and unsatisfactory for use. If this does happen, the bit has to be filed smooth once again. When heating the irons in a flame watch out for a green colour, as this is generally accepted as indicating the correct temperature.

Flux is a very necessary addition to the process as it helps the solder to flow smoothly and protects the surfaces that are to be joined. As with welding, oxidation is a deadly enemy of the soldered joint. Fluxes are available in liquid form, or as a powder or paste, whose melting point is always lower than that of the solder. The final ingredient of course, is the solder itself which is also available in several types. These varieties are mainly determined by the type of work being done. For example there is cored solder which comes impregnated with flux and is particularly suitable for electrical component work done with a small electric iron. The soft solders are based upon a tin/lead

alloy and it is the proportion of tin within the alloy that determines the quality – the higher the percentage of tin, the better the quality – it runs very freely. Steve says that you can actually hear a good solder!. If it is flexed and little cracking sounds can be heard, the solder contains a high percentage of tin which gives it fine qualities. Silver solders are so called because they do contain silver. Often used on window frames, radiators and other brass components on older cars. These solders require much higher temperatures (600-800°C) and require a gas torch.

(Continued)

Chris Graham looks into the theories, methods and applications of both soldering and brazing.

Brazing & Soldering

The process of tinning is the most important stage involved in soldering, and if it is not done correctly then a successful joint will not be possible. To begin with the area that is to be tinned (the iron, the work, or both) must be thoroughly cleaned, and this means to a shiny metal finish. A thin layer of solder has then to be laid down on to the clean surface. This is achieved by applying a mixture of flux and solder to the hot surface, then wiping away any excess with a clean grease free cloth (not a nylon cloth as it will melt), to produce a mirror like finish. Be careful when wiping to avoid rubbing through the skin layer and so breaking the seal, as if this occurs the subject will have to be cleaned again, and the process repeated.

The idea behind the tinning process is to

2
With the two plates tightly positioned together (clamps provide the ideal answer), an extra strip of solder paint should be added along the edge of the joint. At this stage you must ensure that both plates are completely tight against one another. If they are not, un-clamp the plates, clean off the paint, alter the plates as necessary, then start all over again.

3
A domestic blow torch of the DIY paint stripping variety will be quite sufficient as a heat source. The joint should be heated until the solder paint begins to turn silver. This is an indication that the correct temperature has been reached and that the solder is starting to run throughout the joint.

ensure that the solder which is added to produce the bond, is able to run smoothly over all the surfaces involved in the joint. Without this ability it simply will not work. It is wise to 'tin' over at least twice the area that will be needed for the joint, and to waste as little time as possi-

4
With the solder rod in one hand and the blow torch (at a distance) in the other, a run of solder can be laid down along the joint. This should not be too thick but if it is, it should be quickly wiped away with a clean grease free cloth. Further gentle heating will see the excess solder drawn into the joint and if the plates are turned over, oozing out from the other side. By this stage the joint has been well and truly penetrated and should be allowed to cool.

5
Because the heat required for brazing is fairly great, a fire brick is a sensible idea as a base to work on. Obviously the size of the job being undertaken determines the number required and in some cases, a complete hearth will need to be constructed. In this example we chose to braze a stainless steel tube on to a couple of mild steel blocks. It is of course vitally important that all surfaces be thoroughly cleaned, and then the local area can begin to be heated.

ble in between the cleaning and the tinning stages, and between tinning and finishing the job, to minimise the risk of atmospheric contamination. Also, never handle a freshly cleaned sample with bare hands as the grease naturally present in the skin will upset the process. Cleaning in most cases will involve a wire brush and white spirit, but in more severe examples where rust is present, a file may have

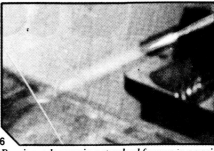

6
Brazing rods come in a standard form or in a variety known as the pre-flux rod. This variety already contains the flux required, which eliminates the time consuming need for the continual dipping into the flux, of the conventional rod. To start with the rod itself (an ordinary brazing rod in this case) was given a short burst from the flame, which was then returned to heating the work.

7
The heated rod is then dipped into the flux (powdered in this case) so that the end becomes coated.

to be used.

The tinned surfaces are then ready to be joined by the addition of more solder. The heated iron is used to melt the fresh solder, and at the same time heat the joint into which the molten solder is to run. A point worth considering here is the type of joint being used. The soldered joint does not provide the most secure of fixings and therefore, it is a good idea to maximise the joint area. The introduction of flanges provides a simple and effective solution, but stepped lap joints can be equally useful. The close proximity of the two surfaces being joined produces an effect known as capillary attraction, by which the molten solder is drawn deep into the joint to seal it fully.

Note that when tinning a surface which is sloping or even vertical it is best to start at the lower end and work upwards thus fully using the heat (which rises).

For our example, we chose to demonstrate a technique known as the sweated joint. This type of joint is best suited to situations where large areas have to be secured, or where the job is too small for the use of a soldering iron at all. The idea is that both surfaces are tinned, then clamped tightly together (it is essential that a precise fit is achieved) and heated, in order that the solder should run. A particularly useful aid to this end is a product known as Solder Paint. It is a liquid that contains both solder and flux, and it neatly achieves the tinning process.

Brazing & Soldering

Brazing

Brazing is a little like the elder brother of soldering. Its results are a lot stronger than soldering, and yet it still relies upon the same basic principles. As with soldering, in brazing neither of the parent metals are melted, a flux is used and the braze is drawn into the joint thanks to capillary attraction. Probably the fundamental difference between the two is that brazing is achieved at a much higher temperature. Brazing can be used to fasten two like or two unlike metals, and this is one of its major advantages, as it makes it versatile should the situation demand it.

8

It is essential that the two parent metals are made hot enough to melt the brazing rod when it is touched on. It is no good melting the rod in the flame and then dabbing it on, as the braze will immediately re-set having achieved nothing. Whilst the braze is being laid down the flame should be played all around the area, not just at one single point. To begin with we found the brick was absorbing too much heat from the metal, which was preventing the necessary level being reached. Simply raising the work off the surface cured this effectively.

9

The flux sets to a black solid on the braze as it cools, but this is easily removed by quenching (cooling the work rapidly in water). In cases such as these where a tube is to be quenched, avoid plunging it straight into a bucket or trough of water as the steam produced can be funnelled up the tube to cause painful burns. The sensible answer is to use a hose.

10

The finished braze having been cleaned provides a reasonably presentable joint.

However, due to the higher temperatures involved (brazing rods melt at around 950°C, soft solders melt at around 200°C), brazing requires the use of a gas torch to produce the heat. But having said that, there is still less heat distortion involved than there is with gas welding, and another advantage is that the final braze is far less brittle than an equivalent weld. This makes it particularly suitable for joints that will have to withstand a certain degree of flexing such as on certain body panels (or bicycle frames). However, it is not advisable to braze car chassis. Our endeavours are illustrated in the picture series. □

MIG WELDING SUPPLEMENT

We take a thorough look at the MIG welder market, test a representative selection and comment on their performance.

MIG welders are not a new development; they have been available to the professional for many years. However, in recent times, the market has been rejuvenated by technological advances coupled with improved mass production techniques. These have combined to create a new generation of MIG welders which are readily available and relatively cheap. It appears that an ever-increasing number of *Practical Classics & Car Restorer* readers are buying their own MIG welding equipment for restoration and repair work. There is a terrific range available now and, in some ways, it could be argued that the consumer is spoilt for choice. The real advances appear to have been made at the middle and lower end of the market which has really opened it up to the DIY user and, consequently, many of you reading this magazine.

Hopefully, this supplement will provide you with a useful insight into the types of MIG welder now available together with information on what they can actually do. In addition, it will also contain the first part in a series of three articles dealing specifically with the successful use of a MIG welder. All aspects will be covered to create a practical guide for the beginner.

The testing

To help us with the assessment of this range of MIG welders we called upon the skills of Bryan Dunmall and a group of his trusty third year motor engineering students at the Mid-Kent College, Maidstone. We asked them to try each welder individually and comment on its performance relating to the quality of the weld it produced and the manner in which it did so. In addition, we were keen to learn their impressions on how easy (or hard) the sets were to assemble and prepare for work and on how much stress the various manufacturers put on the vital issue of safety in the instruction booklets.

The tests devised for the welders were designed to be hard so that a true and useful assessment of performance could be gained. Each was put through its paces and used to make butt welds with samples of metal of varying thickness - 0.8mm, 1.0mm, 1.2mm, 1.6mm, and 3.0mm. Bryan considers that this range represents every thickness that an enthusiast restoring a car was is ever likely to encounter. The finished welds were thoroughly inspected for quality and the best in each case was put forward for an impact test. This was a simple but effective business involving the rounded end of a ball pein hammer. This was aimed directly at the weld and a bowl-shaped indentation of predetermined dimensions was produced in each case. The results were then closely inspected for fracture and breakage.

In presenting the results we thought it best to list the welders in order of their price rather than in a way that reflected our own preferences. Hopefully the results printed here will provide some useful food for thought which anyone currently considering a purchase will find helpful. Obviously the welders shown here do not represent the whole market but we feel that they provide a representative selection.

Maypole No Gas
RRP £199.95

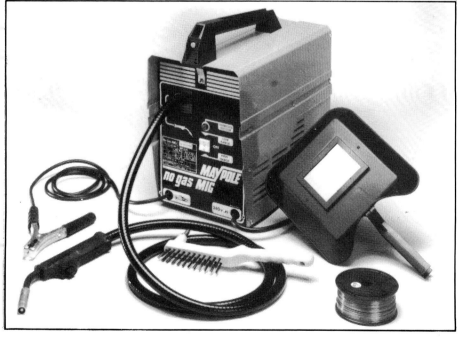

The instruction manual provided brief information on safety together with basic details on setting-up and using the unit. However, a useful fault/remedy chart was included.

This welder represents the 'entry level' of MIG welders. The testers all had problems with excessive spatter (molten droplets of metal thrown out as the weld is made which then cool and solidify on the surrounding area) and they found it difficult to make acceptable welds on the 0.8mm plate. However, as the subject metal became progressively thicker the performance improved. In nearly all cases the appearance of the welds was untidy. The spatter and the burnt flux covering made the welds ugly to look at and considerable final polishing was required to produce something approaching a pleasing result. On the positive side though the welds, in the main, were all strong. Only the joint made on 0.8mm plate actually failed the impact test completely. There were signs of slight cracking in the impact-tested welds made with the 1mm and 1.2mm plate but

those on the thicker samples were up to everything that could be thrown at them!

In conclusion it was considered that the No Gas process provided a basic way of joining two pieces of metal. The results produced are arguably similar to those made by an arc welder although they are many times easier to achieve. It should be noted that this is not a criticism aimed solely at the Maypole No Gas MIG welder and that we feel sure that the other units in this category will produce a similar performance.

became much more acceptable. Although on the thicker samples the appearance of the weld deteriorated somewhat, the strength of the weld produced in each case was excellent. The controls were fine and the gas delivery controlled by the flow valve on top of the cylinder was faultless.

Maypole 100
RRP £249.95

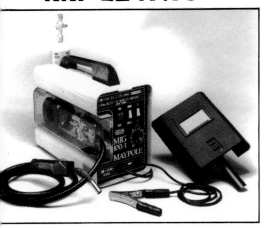

The overall impression of this Maypole unit was good. The instruction booklet, although rather sparse in comparison to some of the others, was adequate. The section on safety was brief but to the point and Maypole include a useful welding fault and possible remedy chart together with details on control settings for various applications. Assembly was clearly covered. Wheels are often considered an advantage on welders but, in this case, they were not necessary. The lightweight construction of the Maypole 100 made it very portable, a very important consideration in a small or crowded workshop.

In common with the other small units tested here the overall performance of this welder was let down by its limited ability successfully to weld thin gauge metal. This was particularly apparent on the 0.8mm samples when the instability of the arc (spitting and popping rather than producing an even crackle) made effective welding very difficult. The testers noted a marked improvement with the 1mm and 1.2mm samples when the arc and the quality of the weld

Clarke Hobby 90
RRP £163.95

The instruction manual provided with this welder was basic. The few lines dealing with safety were hidden within a section entitled

"Over here! There might be a restoration project over here.....!"

'Welding techniques' and could easily be missed.

In use this welder was good. It performed surprisingly well on the thin test samples considering its specification and was generally thought to be some way ahead of the others in its class. On the thicker samples there was further improvement and the strength of all the welds made was excellent.

The impact tests had no destructive effect upon the welds whatsoever. The only slight criticism of the performance was that the welds on the thinner samples were a little untidy. Spatter was evident and final finishing was required for a neat job. No problems were encountered with the gas flow valve which featured a 1-10 scale for ease of operation.

Its only limitation was, once again, its inability to cope with the thin sheet samples. The arc under these conditions was unstable and this was reflected in the poor quality of the resultant weld bead and inadequate penetration. Consequently, the weld produced on the 0.8mm sample was weak and failed the impact test. On the mid-range and thicker samples though things were much better and

the appearance and strength of the welds produced was of a much higher quality. The testers found that this welder was easy to use and that the good instructions made setting-up a simple matter.

Clarke No Gas 100EN Turbo
RRP £182.95

A good instruction booklet was provided with this unit. Details on safety precautions were prominent and clear diagrams illustrated assembly and conversion from 'no gas' to gas function well.

The welder's performance was predictable. On the thin samples the results were poor. The arc was most unstable, there was plenty of spatter and the weld produced was untidy and very weak. Things improved as the thickness increased and reached a peak with the 1.2mm sheet steel. In this case the arc was good, the weld produced was quite clean and uneffected by the impact test. Generally though, spatter was a problem throughout the test. On the thicker metals the testers found it hard work to make a weld although, when they managed it, the strength was good. No problems were encountered with the controls or the gas valve. This unit was fitted with a turbo fan which worked well and there were no problems with cut-out.

Bryan concluded by adding that the no gas system, although being somewhat crude, produces very solid results on thicker metal sections. This makes it ideal for work on chassis rails etc. He went on to say that with a conventional gas MIG welder a beginner can sometimes be fooled into thinking that a strong weld has been made. In reality, though, what has really happened is that the line of welding bead has not penetrated down into the metal sufficiently to create a strong join although it looks perfect from the top. This risk is considerably lessened by using a no gas unit as it tends to 'blast through anything'!

Clarke 100E Mk2
RRP £187.95

The instruction booklet supplied with this machine was of a good standard. Information on safety was prominently located on the first page and set out clearly the 'dos and don'ts' of the process. The remaining 24 pages contain all the necessary information relating to assembly, welding technique, specifications, setting guides and fault finding

The 100E Mk2 proved to be a useful tool.

Clarke 120E
RRP £235.95

This welder was the best of the selection sent by Clarke. One of the reasons that it stood out from the others was because it was fitted with wheels at the back and a long handle protruding from the front making it very easy to

wheel about. There was some muttering from the testers that the long handle may get in the way when welding in tight corners and some thought bumping one's head on it was a painful possibility. However, overall the handle was considered a worthwhile attachment.

The instruction manual was almost identical to the others supplied in the Clarke range but, in use, the 120E was well ahead of its stablemates. It was able to cope well with the thin metal and produced good results that were neat and strong. Every weld made passed the impact test with flying colours. The best welds were made on the 1mm, 1.2mm and 1.6mm test samples but the quality, not the strength, dropped away a little on the thickest sample. The control layout was straightforward and no problems were encountered in this respect.

Maypole Dual MIG 120
RRP £249.95

The instruction booklet provided with this unit left something to be desired. It was printed in four languages and, once we found the English section, it became apparent that there were no details on the safety precautions required. Much of the basic setting-up information was printed only in Italian which wasn't much use although there were diagrams to help with the explanations.

This welder can be used either with or without gas. However, because the set was inadvertently supplied without gas, we could only test it in the less effective no-gas mode using the special flux-cored wire. Not surprisingly, the results were rather disappointing. Its performance on the thin gauge samples was very poor and it improved only slightly as the thickness increased. The arc condition got better on the thicker samples but, unfortunately, the weld quality did not match this. The best results were obtained on the 1.2mm and 1.6mm samples and the strength of these two welds was very good. Spatter was a problem in all cases. This was especially so in the case of the 3.0mm sample although the weld produced was very strong and performed without fault in the impact test.

lems with over-heating. In use the testers found some difficulty in making a successful weld on the thinnest gauge sample. The arc was rather unstable and spatter was a problem. However, above the 1mm thickness the quality of the arc — and so the ease of welding — improved steadily. On the 3mm plate the arc was very good and a satisfactory weld was produced although the penetration was questioned. Overall it was thought that this welder performed best with the mid-range thicknesses when a stable arc led to neat welding with good penetration. The strength of the welds produced improved consistently as the subject metal thickness increased. The welds on the thinnest samples were weak but there were no such problems on the thicker ones.

SIP Migmate
150 Turbo
RRP £353.00

This welder was supplied to us with 1mm heavy gauge wire which rather limited it in the welding tests. The thickness of this wire made the welding of thin metal plate impossible and so the testing was restricted to the thicker samples. However, it produced good results of a very high quality. The tests, which centred around the 1.6mm and 3.0mm samples only, proved the ability of the welder on these thicknesses. The welds produced passed the impact test with ease and, consequently, the testers felt sure that the 150 Turbo's performance on thinner material was likely to be equally encouraging.

This unit was the only one of the SIP welders supplied to us with a flowmeter type gas regulator valve. The instructions advise that this be set to about '3' when welding begins but there is no indication why this should be or when it should be set differently. Every control worked well on this unit and it was enthusiastically received by all the testers.

SIP Migmate Super
RRP £273.00

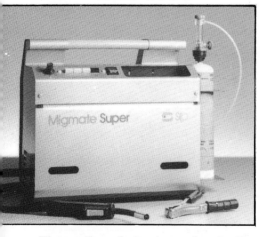

The quality of the instruction booklet supplied with the Migmate Super was every bit as good as that described for the Handymig; SIP sensibly produce information of the same standard and degree of usefulness for every one of their models.

In use the testers experienced difficulty in achieving a stable arc at the lower settings with this unit. This made welding the thin plate difficult. The situation improved as the thickness of the subject metal was increased and the quality of the weld matched this too. Strength, in most cases, was good although the joint made on the 0.8mm sample showed some slight evidence of cracking after the impact test. Unfortunately, during the welding of the 3mm sample, the welder overheated and this triggered its thermal cut-out mechanism. This is a safety device which most MIG welders feature nowadays. In fairness the welder was being worked hard; however, it was the only one to cut out under test and it happened after about 25 minutes of work. It took about 45 minutes for it to cool sufficiently to allow work to re-start.

Another slight problem that the testers found concerned the gas valve. Apart from that fitted to the 150 Turbo model described later, these SIP welders all featured a very simple on/off type design which proved rather unreliable. In one case the pressure relief valve was leaking all the time so gas was being wasted. In another the valve was difficult to attach to the bottle so that the gas would flow correctly — it could be screwed into place with ease but no gas would flow. In the instruction manual you are advised repeatedly to remove and replace the valve to overcome any flow problems and, in one case, a tester spent 20 minutes trying to get this to work. Two out of the three valves supplied gave problems. Overall though, this unit proved itself well and received much praise for its design and ease of use.

SIP Migmate 130 Turbo
RRP £313.00

One of the first points noted about this model is that it features wheels, making it easy to move about in the workshop. This made it an instant hit with the testers!

The instruction leaflet equalled the other SIP units and the welder itself was simple to set-up ready for welding. This welder worked well overall and there were no prob-

Migatronic Rally 166
RRP £445.00

All three Migatronic welders in the group came with extremely comprehensive instructions and Migatronic were the only supplier

to send a representative with the machine to demonstrate their operation. The Rally 166 was the cheapest Migatronic and the testers liked the controls and found the machine easy to set up. It welded well on all metal thicknesses except the top 3mm range where the weld failed to penetrate properly. Apart from this the only criticism was that the torch trigger's position wasn't particularly good. Bryan's overall comment was that this machine represented good value for money, especially in view of Migatronic's back-up service.

Clarke Professional 185E
RRP £519.95

Our testers found it hard to produce a satisfactory weld on any metal thickness, but particularly at the lower end of the range, without 'pulsing' the trigger. This made welding much harder than with any of the other machines tested, though once the technique was mastered the results were comparable with the others in the group. One of the testers (each machine was tested by three people, selected to represent a cross-section of abilities) disliked the handle. Instructions supplied with the machine were very basic

(two A4 sheets in four languages), the only reference to safety matters being a small section of 'advises' tucked away on page 3. Still, it was better than any of the DIY machines, as would be expected.

Migatronic 180MX Typical price £848.94

As would be expected, this machine was a little more complicated than the Rally 166, but still exceptionally easy to use and understand. Its performance across the entire range of thickness just could not be faulted, all welds looked good and penetrated the entire thickness. A good feature of this machine was that it still produced results (though obviously not as good) when on the wrong setting for the thickness of metal. Bryan commented 'Definitely an upmarket, professional machine'.

SIP Autoplus 200 RRP £819.00

This machine had two setting controls, a coarse and fine adjustment, giving a total of

twelve different settings for different metal thicknesses. It performed well across the entire range producing good quality welds — our testers could find little to criticise about the welds' appearance and nothing at all about their strength. The only real criticism was that the machine wasn't particularly forgiving of incorrect setting; it had to be spot on before the Autoplus would give of its best. Instructions supplied were comprehensive enough and we liked the way that the safety precautions were first in the book.

Maypole Pro Turbo 170 RRP £499.95

Our testers found this machine straightforward to set up and very easy to use in the main, particularly on lowest and highest settings for thinnest and thickest. Once the technique was mastered it produced good results in terms of appearance and penetration right across the range. No complaints on the strength of finished welds. One major grumble was that the Maypole continued feeding wire after the trigger was released, wasting wire.

Migatronic 185 Typical price £1,526.00

This machine is basically an Automig 180 with panel shrinking and stud welding (for panel pulling) facilities. Whether or not it is worth the extra money will depend on how often these facilities are likely to be needed; performed exactly the same as the 180.

Conclusions

We were a little concerned when we undertook this supplement that the differences between most of these welders would be hard to determine. However, experience has proved us wrong and we feel that the exercise has highlighted not only some worthwhile general trends but also a few individual shortcomings.

We will leave the latter for you to judge but, in terms of the general trends, we can offer the following thoughts. Quite clearly the majority of MIG welders in the middle and lower end of the market are not yet at a standard which allows them to confidently tackle work on thin metal panels. Many of

the manufacturers claim that they are but, judging by what we have found, some of the advertising brochures are perhaps a little optimistic. If highly experienced and skilled welders like Bryan Dunmall find using them for this type of work demanding then, quite frankly, what chance does the newcomer have of success? The secret is to appreciate the limitation of the machine and not to expect too much.

There is no question that the DIY type MIG welders have come on immensely in the past few years and some would argue that they are now matching the standard that was being produced by the professional equipment not so very long ago.

The No Gas sets are a new development and there is clearly room for more improve-

ment yet. The special flux-cored wire required for this type of operation is quite expensive (roughly about twice the price of normal welding wire) and a standard size reel will cost about the same as one of the small gas cylinders used on the conventional MIG welders. These cylinders last for varying lengths of time and this depends, to a certain extent, upon the contents. A cylinder containing pure carbon dioxide will run for about an hour while one containing a mixture of carbon dioxide and argon is likely to be used up in about 10-12 minutes.

However, in the final analysis, as in so many cases relating to tools and equipment, it all comes down to money. The hard fact is the more money you spend the better will be the result. □

On test:
The SIP Handymig/Handymig Gasless

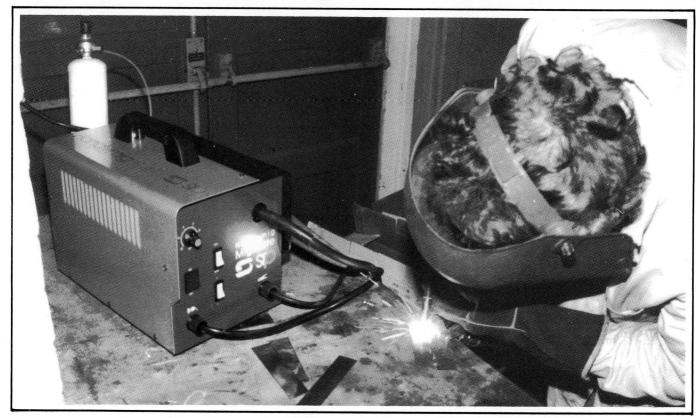

Chris Graham and Bryan Dunmall review these two interesting stablemates from SIP

As we discovered a couple of months ago in our extensive MIG welding supplement (December '88), competition among the manufacturers of the smaller sets is very hot. Reputations count for a great deal with the buying public which is why we were anxious to make our testing procedures authentic. Unfortunately, one of casualties of that exercise was SIP's Handymig which suffered a motor failure that prevented a full assessment. SIP were understandably rather disappointed with this result and offered us the chance to run a re-test of the Handymig and, in addition, to take a look at their Handymig Gasless model which was not available for testing before.

Bryan's first reaction to the Handymig was very encouraging. He finished the first weld and lifted his head shield to reveal a big grin! "This one welds beautifully" he said. He experimented by making butt joints with a range of metal thicknesses and found no problems whatsoever. Admittedly care was needed with the thin samples but this was only to be expected. One of the findings in the supplement had been that the smaller units all had their limitations when it came to working on thin gauge plate. The arc remained stable on all settings which made the set easy to manage and the wire feed control was smooth and efficient. The unit per-

The Handymig Gasless is identical to the Handymig model apart, of course, from the absence of a gas bottle. The controls on both are the same and access to the wire reel and tensioning mechanism is gained by removing the smoked plastic cover seen on the right.

Welding with the Handymig Gasless proved to be something of a revelation. Bryan was surprised by the quality achieved in comparison with some of the others which he has tried recently.

formed well on the thicker gauges of metal and Bryan, together with the other members of the staff at the college who tried it, was very impressed with the quality of the welds produced. SIP quote the Handymig as having a list price of £172 but they were keen to point out that, by shopping around, those interested in buying one will be able to secure a considerable reduction on this price.

The SIP Handymig Gasless, which has only recently been introduced, proved to be quite a revelation. Initially Bryan imagined that this machine would perform in much the same way as the others that we had already tested. He was expecting a somewhat crude result with plenty of heat generated but with little control. However, he was surprised. The arc produced was stable and did not deposit excessive spatter on the surrounding metal. The colour of the arc was noticeably whiter than usual and we noticed that heavy deposits of light-coloured powder were left on the surface. This, however, was nothing more serious than the residue from the burnt flux-cored wire used for this type of welding and it was easily removed. After finishing his testing he considered this unit to be the best of its type that he had seen. Its list price of £145 pitches it close to its conventional rivals but, once again, shopping around will certainly result in considerable savings. Finally, we are interested to note that SIP have modified the regulator valve fitted to the Migmate Super and this provides a great improvement. □

For further information ring SIP Helpline (0509) 503141.

SMOOTH OPERATOR

It's always encouraging to test a brand new product and find that it really does its job well. There are so many disappointments around these days that we become almost conditioned to expect failure before success. Happily, however, this was not the case on the occasion when Bryan Dunmall got his eager hands on the new BOC Migmaster 130 Turbo, recently featured on our *Goody Corner* pages.

BOC are probably best-known to most of you for their work with gases and, indeed, this new welder is designed to be used with their extensive range of products. The company are constantly working on new projects and ideas and one of these, which has just come on sale at the 70 BOC Cylinder Centres up and down the country, is Argoshield TC (total capability) gas. But the virtues of this will be discussed later.

On the face of it the Migmaster 130 Turbo is just another MIG welder – it looks essentially like all the rest. The controls, four separate voltage settings and a stepless wire speed adjustment (calibrated 1 to 10), are simple. A side panel is removed to fit or change the filler wire reel, and wheels plus an extension handle are also supplied.

The instruction booklet provided is well-produced and comprehensive. Safety is given pride of place at the front and sections on setting-up, operation, basic welding techniques and typical control settings follow on afterwards. Bryan was impressed with the 'control setting suggestion tables' and found them useful as a starting point. However, he thought it important to note that they related to the use of TC gas for mild and stainless steel and argon for aluminium.

Phil Barber, one of BOC's expert demonstrators, gave a dazzling display of the welder's capabilities and then it was left to Bryan to form his impressions.

The Migmaster was set-up with BOC Argoshield gas (a mixture of argon and carbon dioxide) and, once Bryan had familiarised himself with the settings, there was no stopping him! He concluded that its performance on mild and stainless steel was excellent. He liked the torch and thought that the wire feed was of above average quality – the adjustment was very sensitive, giving great flexibility and control.

He made welds on all thicknesses of material from 1.2mm to 3mm and could not fault the results. However, the big test was yet to come. The manufacturers claim that this unit is capable of successfully welding aluminium. Bryan told me that this was one particular aspect of the manufacturers' claims that he was keen to test out.

A reel of aluminium/5% magnesium wire was loaded, a cylinder of pure argon connected and away he went. It worked like magic to produce a clean and relatively tidy, strong weld. There were no problems of weakening caused by annealing, as you are likely to get when gas welding aluminium. Bryan was amazed.

We test BOC's new Migmaster 130 Turbo welder – Chris Graham brings us his report.

According to BOC the Migmaster's secret, in this respect, is the high power output at the top end of its working range. BOC paid great attention to the other welders on the market and tested each one thoroughly for performance and power output. Many fell sadly short of the manufacturers' claims. This short-fall poses all sorts of problems when working with aluminium.

To explain this fully it is necessary to understand the two mechanisms by which a MIG weld can be made. When you are working with steel the filler wire from the handset is literally pushed into the molten weld pool where it melts to make the joint. This is known as 'dip transfer' welding and it is quite suitable when using steel filler wire. However, if you need to weld an aluminium panel or casting then you must select an aluminium-based filler wire. The properties of aluminium mean that, as it approaches its melting point, the metal becomes very weak. Trying to operate with this wire using the dip transfer method will lead to a very unstable

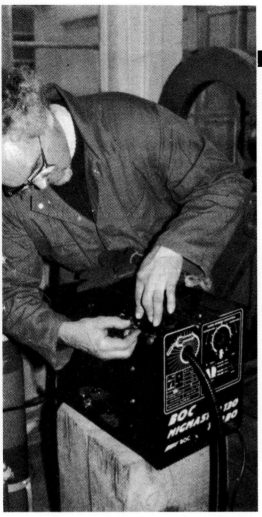

Bryan was impressed with the wire feed mechanism and its tension adjustment system. The filler wire reel is held in place with a simple push-on plastic clip washer.

▶ *The Migmaster 130 Turbo will produce top quality results when used with BOC gas. Three different sizes of cylinder are available; the one illustrated is known as the 'B' size and contains about 3.5 cubic metres of gas which will provide 1.84 hours of continuous welding. It costs £17.25 per year to rent the cylinder and £15.87 TC, £20.91 Argon for each refill.*

In use the Migmaster was everything that BOC had claimed. One novel feature of the handset is that, when the trigger is pulled, the gas is switched on an instant before the power. This purges the immediate atmosphere of oxygen to create the perfect, impurity-free conditions in which the arc can strike.

The results speak for themselves! This is the quality of weld possible on steel using BOC's TC gas. I have never seen a smoother bead laid down straight from the gun.

The Migmaster's performance on aluminium was startlingly good for a welder of this size. BOC put this success down to an improved and extra efficient transformer which produces all the power that it should.

Bryan gave the aluminium weld a scientific test for strength which it passed with flying colours!

arc – the tip of the wire literally will keep dropping off before it has been melted fully and, consequently, there will be poor flow and a lack of control.

The answer to this problem is 'spray transfer' welding. To achieve this the welder is turned up to a high power setting (assuming, of course, that it is capable, in reality, of achieving such a setting!) so that a long arc is created. This will prevent the wire from actually touching the surface and the increased power will ensure that it is melted completely. The 'liquid' metal is carried down on to the surface in a fine spray within the arc, giving great control.

The Migmaster has the necessary power to achieve an acceptable degree of spray transfer and Bryan found that it could weld 1.6mm aluminium – the normal sheet metal thickness used for aluminium panels – comfortably. Quite simply he found the Migmaster's performance on aluminium exceptional!

Although the Migmaster itself represents a considerable achievement it should not be forgotten that a good proportion of its success results from the BOC gas used with it. The new TC formula represents a particularly interesting product. Up until now, users have only had the choice between Argoshield 5 or 20 (the figures relate to the percentage of carbon dioxide mixed with the argon). Generally speaking, the quietest and most controllable arc will be obtained when using gas with a high proportion of argon (low CO_2 content – Argoshield 5). This is ideal when working on very light gauge metal because it helps to prevent burn-through.

Higher proportions of carbon dioxide (Argoshield 20) produce more heat and a fierce arc which makes it better suited to welding heavier gauge material. These two represent effectively the two ends of the common welding spectrum. Conveniently, BOC's TC gas falls in between these two alternatives in terms of CO_2 content. It can be used very successfully on both thick and thin material. The arc is controllable but produces sufficient heat to work well at both ends of the range and it produces a superbly neat weld.

A common problem with the smaller MIG units is that their thermal cut-out mechanisms – which cut off the power when the machine becomes over-heated – take too long to recover. Some we have tested in the past took up to half-an-hour before welding could start again. BOC claim that the Migmaster will recover in about five minutes, which is very fast. Unfortunately, we were unable to test this out because we couldn't get the machine to over-heat in the first place!

In conclusion Bryan feels sure that this new welder will be a success. Retailing at about £200 (welder, handset, regulator, gas hose) it is certain to set the cat among the pigeons on the marketing front. It has the performance to match that of much larger and more expensive machinery and its abilities with aluminium proved to be a revelation. For more general information (including the whereabouts of your nearest BOC Cylinder Centre) you should contact Donna Evans or Diane Brown on 0483 579857. □

Last month, as a competition prize we offered a Migatronic 5000MX Mig-Welder, donated by The Welding Centre. We arranged for Colin Ford, a local car restoration specialist to try out the Migatronic and to say that Colin was impressed is something of an understatement.

MIGATRONIC 5000MX WELDER TEST

The Danish-made Migatronic is designed as a logical 'next step up' from a Mini-Mig for either the serious car enthusiast or the professional user, by giving the performance one would expect from a professional machine but at a fraction of the cost. It operates from a single phase 240V AC supply, current output varies from 25 to 150 amps and the open circuit voltage varies from 20 to 32 volts. It is sturdily constructed, has clear and easy to operate controls, and comes complete with a Migatronic 120D torch with swivel head. Spot, stitch and continuous weld functions are available, and the Migatronic takes wire sizes 0.6mm and 0.8mm. The unit is supported on two wheels and two castors, is easy to move about, and there is the possibility of fitting a larger gas bottle without impairing the unit's portability – important consideration for the serious or professional user!

At the time of writing Colin has been using the Migatronic for about a fortnight, during

The Migatronic is sturdily built within compact external dimensions, and, as explained in the text, it would be perfectly possible to fit a larger gas bottle than that shown here without restricting the unit's portability too much.

All controls are clearly marked and laid out for ease of use.

The Migatronic 5000MX in action. During the fortnight's test period the Migatronic was used to weld things as diverse as a washing machine case and the Triumph TR6 that Colin is working on here.

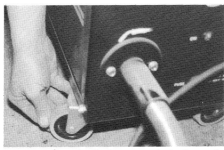

The castors and wheels are particularly sturdy and the whole machine is clearly designed to provide many years of use.

which time his regular welding equipment has been gathering dust in the corner of his workshop. The 5000MX has been used a great deal in that time and, apart from the usual workshop duties, it has been used successfully to repair Colin's washing machine! It coped well on steel of a variety of types and thicknesses. What particularly impressed Colin, however, was the Migatronic's durability. He pointed out that things like the wheels were particularly well made, a minor point maybe, but nonetheless typical of the attention to detail that has gone into the Migatronic's design and construction.

The Migatronic 5000MX is available from **The Welding Centre, 293 Ewell Road, Surbiton, Surrey. Tel: 01-399 2448/9.** It is priced at £399.00 plus VAT and, if applicable, carriage. ☐

DIY TIG WELDERS

A new generation of DIY welders

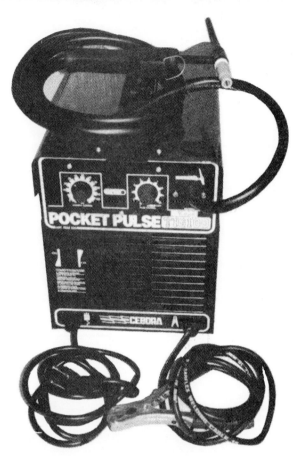

Practical Classics & Car Restorer were in at the beginning of the MIG welding revolution (and at the forefront of encouraging it), when pocket MIG welders first brought this type of welding within the reach of the DIY enthusiast. Now with the introduction of pocket TIG welders we believe that *they* could be the next generation of DIY welders.

Until now TIG welders have sold at £1,000 and upwards, restricting them to professional use. The new 'pocket' TIG welders however are available at prices between £350-£400 making them a real possibility for the more serious DIY welder (although they cannot weld aluminium which the expensive models can).

SO WHAT ARE THE ADVANTAGES OF TIG?

TIG welders offer all the advantages of MIG welding, *plus* many of the advantages of gas welding, without the disadvantages.

TIG welders use an electrical flame conducted by a tungsten tip and shrouded by argon gas. They operate on a similar principle to gas but without the heat distortion that gas causes and without the inherent dangers of a big gas flame. It is also more 'user friendly' than gas welding.

TIG can weld thinner metals than MIG, without burning through, by virtue of it being used at much lower currents and therefore has a lower heat input.

When welding on a gap the TIG's arc goes straight to the metal and doesn't shoot through the gap (as gas welders do). A standard gas welding filler rod can be used with a TIG (in the conventional way) when welding a gap. When welding two pieces of metal butted together a perfect weld can be achieved without the use of filler wire.

TIG welding gives a far neater weld than MIG - as good as any gas weld - without a slag build-up. In fact the clean tidy weld and lack of slag build-up can give you a weld that is almost invisible.

AVAILABLE EQUIPMENT:

Cebora Pocket Pulse TIG100. Manufactured by the Cebora company of Italy. The torch head is British manufactured and uses British accessories. This TIG welder is DC, controllable from 5 amps up to 100 amps and is ideal for mild steel and stainless steel (not aluminium) and can be used for copper and cast-iron. It features an electronic 'soft start' system which provides easier starting and prolongs the tip life of the tungsten. Supplied complete with flow meter, one bottle of argon gas and one tungsten tip (but without headshield). Available, or further information from, **Surrey Welding Centre,** 293 Ewell Road, Surbiton, Surrey (Tel: 081-399 2448) or **Transpeed Limited,** 311 Portland Road, Hove, East Sussex (Tel: 0273 774578).

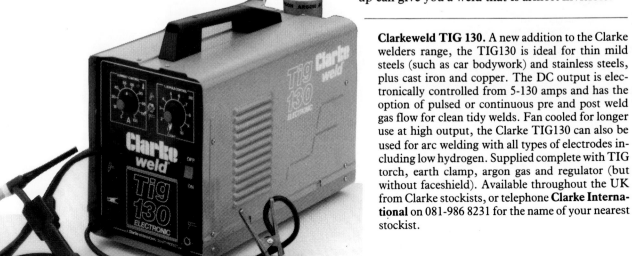

Clarkeweld TIG 130. A new addition to the Clarke welders range, the TIG130 is ideal for thin mild steels (such as car bodywork) and stainless steels, plus cast iron and copper. The DC output is electronically controlled from 5-130 amps and has the option of pulsed or continuous pre and post weld gas flow for clean tidy welds. Fan cooled for longer use at high output, the Clarke TIG130 can also be used for arc welding with all types of electrodes including low hydrogen. Supplied complete with TIG torch, earth clamp, argon gas and regulator (but without faceshield). Available throughout the UK from Clarke stockists, or telephone **Clarke International** on 081-986 8231 for the name of your nearest stockist.

Kel Arc BODY WELDER

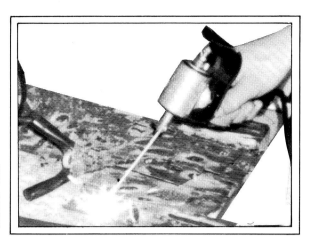

John Williams blows the dust off his arc welder and tries out a tool which revives the use of electric arc welders in DIY restoration.

From time to time over the past year or more readers have sought our opinion of the Kel Arc Body Welder. This is a tool which is used in conjunction with an arc welder but which, unlike an arc welder on its own, is said to be particularly suitable for welding the thin sheet metal associated with body panels. It has already made a tremendous impact in America, where it is currently outselling the DIY MIG welders. Does it work?

If you have tried to weld body panels or other thin sheet metal using an arc welder you will be familiar with the difficulties. You will know, for example, that the best welding you will do for minutes at a time may well be when the welding rod sticks to the workpiece while you are trying to strike the arc. Then, when you *do* get going it seems impossible to weld in a nice neat line as the rod is so flexible

The end of the cable from the Body Welder simply clips into the arc welder's electrode holder and the arc welder's earth lead is attached (by a large crocodile clip) to the workpiece as usual.

The Kel Arc Body Welder enabled me to weld thin pieces of sheet steel in a way which would not have been possible with the arc welder alone. It is worth getting plenty of practice on scrap metal before tackling actual body panels.

(certainly until half of is used up) that it is virtually uncontrollable, yet there seems no prospect of improving on the erratic bead which you have made without an immediate risk of burning a hole right through the sheet metal. There seems to be hardly any margin for error. Either you are leaving an impressive pile of slag but failing to penetrate the metal with the weldpool (because you are working too fast and/or not developing enough heat), or you are making welds but a lot of holes as well.

The Kel Arc Body Welder can solve these problems, not automatically, but I have found that it *does* make it relatively easy to eliminate such difficulties. As its name implies, arc welding depends for the source of heat on the electrical arc between the welding rod and the workpiece, both being in very close proximity to each other. The arc cannot occur if the rod is actually touching the workpiece or if it is too far away from it. Whereas when using an arc welder in the ordinary way it is the operator who has to control the gap between the tip of the welding rod and the workpiece, when the Kel Arc Body Welder is in use the welding rod can be rested against the workpiece and the arc is created by the pulsating action of the Body Welder in which the arc is interrupted, giving a 'stitch-weld' effect. Furthermore, the Body Welder effectively reduces the output of the arc welder and provided that it is adjusted correctly the welding action is much less harsh than that of most arc welders on their lowest settings.

The Body Welder is supplied with detailed instructions which recommend that the beginner should take the time to obtain plenty of practice on pieces of scrap metal. As it is a long time since I last tried electrical welding I did just that and it was well worth the effort. At first, presumably because I had been more accustomed to an arc welder, I found that I was working too fast and not getting sufficient penetration of the sheet metal. It was easy enough to adjust my technique until I was producing strong welds.

Yes, the Kel Arc Body Welder works well and, what is more, it does not cost a fortune. You will need an arc welder on which it is possible to vary the output settings down to 50amps and preferably a little lower but these can be bought for as little as £40-£50. The Body Welder itself costs £24.95 and you will need welding gauntlets, rods and a face mask if you do not already have one. So even if you have to buy an arc welder as well, for less than £75 you could be ready to start welding body panels.

Kel Arc Kompact Welder

A hand-held arc welder which offers a different approach in welding technology and has achieved a unit as easy to use as an electric drill.

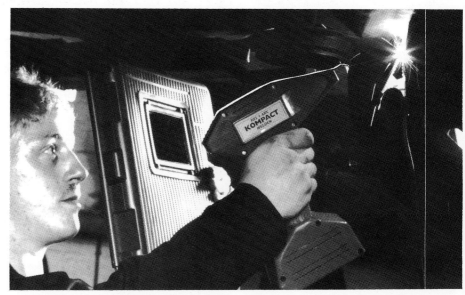

This unique tool was first seen on TV's "Tomorrow's World" and is the first hand-held arc welder. It uses FET (field effect transistor) to eliminate the need for a bulky transformer. In fact all the 'works' of the unit are built into the bottom end of the handle.

It had been designed to be operated by the novice and comes complete with a carrying case which has a lid that doubles for a face shield. The price of around £90 puts it in competition with the larger bulkier conventional arc welders and well below the price of a MIG welder.

A neat design feature is the lid of the carrying case that can be removed to become a welding face shield.

The Kel Arc Kompact Welder comes complete with carrying case, electrodes and striking block (to aid with the striking of an arc) and full, well-detailed instructions.

OUR IMPRESSIONS

The problem with arc welders is the difficulty that the inexperienced user has in striking an arc and then, having struck an arc, having the rod attach itself to the job (or 'stick-with-it-welding' as we call it). The manufacturers claim that this unit is easier to use. Our own initial experiences of using it were on a hot Summer's day. It overheated quite quickly, even before a decent arc or run of welding could be started. It then took some considerable time before it cooled down enough to be used again. We next tried using it on a cooler day and we were rewarded with better results but it was still not capable of very much continuous welding before overheating caused it to cut out.

However since then the FET has been redesigned to allow for far better cooling and it can now cope with more continuous use. It is still an arc welder though, and not everyone enjoys great success with arc welders. To help with this Keller include their own 'striking block' which helps to ensure easier arc striking and to eliminate rods sticking.

The unit balances nicely in the hand and, with the assistance of an extension lead, is extremely mobile, allowing rapid access from one part of a vehicle to another.

The Kompact Welder is only suitable for sheet metal and is not meant for thicker metals. It is therefore unfair to expect a hand-held arc welder of this nature to tackle major work and it is only suitable for the numerous smaller bodywork welding jobs encountered in the course of a restoration.

Further information available from: T F Keller & Son, 24 Cattle Market Street, Norwich NR1 3D2 (T. 0603 624681).

SpotWelding and SpotWelders

S pot welding is simply another type of electric welding. Whereas during normal arc welding the current flows between the welding rod (electrode) and the workpiece itself, creating an arc and fusing the metal, during spot-welding the current flows between two electrodes both attached to the machine. The workpiece is inserted between these, and is brought up to melting point almost instantaneously when the current flows, the two surfaces melting together directly under the electrodes. The electrodes also assist the progress by squeezing the two (or sometimes more) sheets of metal together as the machine is operated.

There are two big advantages on the side of spot-welding: speed, and minimal distortion. It is much quicker to fasten a four-foot sill in place using a run of spot welds, for instance, than it would be using gas or arc welding equipment. Distortion is less with a spot-welder because the heat created is very local and applied for only a very brief period.

The spot-welder itself is a simple and robust piece of equipment consisting (basically) of a heavy-duty coil which produces the amperage necessary, and a pair of detachable arms which hold the electrodes. As the electrodes are brought to bear the current is switched on and the weld made. Some spot-welders have automatic timers controlling the duration of this, pre-set according to the thickness of the metal being welded. Mostly however, especially with the simpler DIY machines, the operator judges the time required – it's usually a second or so.

Using the spot-welder is equally simple, though of course like everything else it takes a bit of practice to become proficient. There are two golden rules when it comes to spot-

Ought you to buy a spot welder? Paul Skilleter goes into spot-welding, what it is and why it is used extensively in the car repair business.

welding – the metal surfaces must be in flat, close contact, and there must be no paint or rust scale on either surface. Bright metal only is permissible. This means cleaning the surfaces prior to welding with a grinding disc or similar, and dressing the seams to be joined flat with hammer and dolly. You shouldn't rely on the clamping action of the spot welder's arms alone either, if there is a suggestion of the metal springing apart, but employ clamps at strategic points to ensure good contact.

This is really all standard welding practice, and should be followed in other areas too. For instance, when attaching a wing don't start at one end of the engine bay flange and keep going until you reach the bulkhead – having lined up the panel correctly and clamped it in place, secure it with a spot weld at each end first. Then complete the line of spot welds. As for how many welds to use, no matter how careful you are there is an inherent risk of failure in the process so more rather than less is the rule here – on average a weld every inch/inch and a half is sufficient for securing a sill.

It's an advantage to position the welds even closer on other jobs. For example, the spot welder is ideal for securing door repair panels – you could never weld across a door skin with other types of equipment, but you can with a spot welder thanks to the lack of distortion. In these circumstances you can build up the spot welds so that they are almost continuous, as this brings the overlapping skins closer together – making subsequent filling easier and minimising the risk of cracks due to flexing appearing later on. The same lack of distortion means you can also make up and fit patches to wings, where the application of gas would make a real mess of things.

Not that it's always plain sailing – the beginner may well find that at first he does nothing but burn holes right through the metal, which is simply due to holding the current on for too long. Weak spot welds

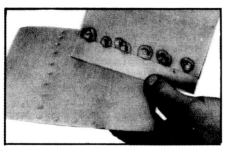

Spot-welding produces very neat results if carried out correctly, as shown on left. A beginner usually ends up with the blotches shown on right however!

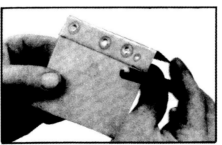

Holding the current on too long can punch a hole straight through the metal, as shown here. Metal is then also deposited on the electrode which must be cleaned.

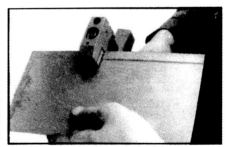

Surfaces to be joined must be bright metal, and in close contact. A swaging tool like this costs around £30 and the 'step' it makes allows two panels to be joined flush.

SpotWelding and SpotWelders

This is ideal for partially re-skinning a door – here the stepped panels are being spot-welded together on our Volkswagen. Gas welding would have produced unacceptable distortion.

which break as soon as you look at them can be another problem, and this is caused by the metals of the two surfaces not fusing together – proper fusion is the key-note of all welding processes. Practice is the answer and if you are a beginner, cut up lengths of mild steel and get in plenty of practice, testing the welds for penetration and strength by trying to force the two strips of metal apart afterwards, using pliers or even a hammer and chisel. You'll soon get to know what your equipment can do, and what constitutes a good weld.

Maintaining a spot welder is easy, but shouldn't be neglected. Most particularly,

An alternative to the conventional spot-welder is the attachment available as an accessory to an arc welder. A central carbon is lowered and an arc briefly created to make a spot weld.

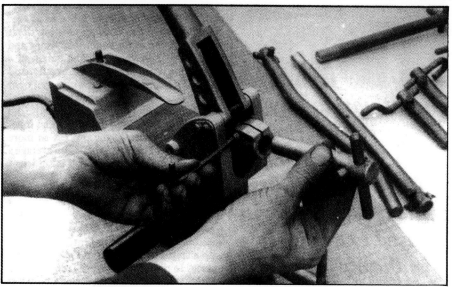

A wide selection of arms are available to enable you to get round door pillars and so forth; they are quickly changed using an Allen key. Electrodes are secured with a wedge.

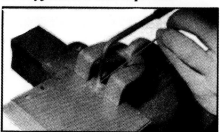

Plug welding: a hole is drilled through the surfaces to be joined, then filled with weld.

always ensure that the electrodes (or tips) are correctly shaped (a tool is provided for restoring their correct contours), and free from deposited metal – which occurs if the tips are overheated, and leads to even more burning, lots of sparks and more holes. Variously shaped arms are available so that the welder can be inserted round door pillars and the like, and a collection of these can be built up if you don't buy the whole outfit from the start.

Spot welders aren't incredibly cheap, so you will have to judge whether you'll use it enough to make its purchase worthwhile. They start at around £130, although an automatic one will cost upwards of £180. A full set of bodywork arms will set you back around £90.

If you already have an arc welder, a 'one-sided' spot welding attachment might be worth looking at. These usually consist of a type of gun having several prongs which are pressed against the workpiece. An internal carbon rod is then brought down and an arc struck, this fusing the metal. The results are not as neat as with the 'proper' spot welder but are serviceable – and of course the initial cost is far less.

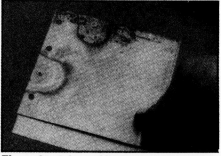

The results of plug-welding are neat if carried out expertly, but as can be seen the problem of heat-spread remains so the process is not really a substitute for electric spot-welding.

A similar effect to spot welding can be obtained with a gas torch as shown in the pictures. This simply involves drilling through the surfaces to be joined, clamping them tightly, and back-filling with the welding rod. An expert can produce a very neat 'spot' weld (actually termed a plug weld) using this method, though even he cannot overcome the extra heat produced by the gas flame.

So, that's spot welding. The money is undoubtedly well spent if speed is important to you, or you want to faithfully reproduce the way the manufacturer of your car originally put it together. It also allows you to tackle repairs to doors and wings that might otherwise be impossible. You pays your money... □

Maintaining the electrodes' correct profile is essential for good welds – they soon become flattened. This tool does the job and fits to an electric drill.

YOU TOO CAN GAS WELD

PART 1 OF A 2 PART FEATURE

Gas welding is a skill — it takes practice and experience. In this first part of our two-part feature Michael Brisby describes the equipment you will need and guides you through the first stages of learning how to use it.

Gas welding is a bit like learning to drive a car — you can learn all the theory, but that knowledge is little better than useless without practical experience and proper instruction. There is no chance of getting the welding equipment home on Friday and being able to tell an admiring audience on Sunday evening that you are now a welder.

DEFINITIONS

Conventional gas welding involves a torch in which a mixture of oxygen and acetylene are burnt to melt metals. Oxygen and acetylene are both colourless, but while oxygen does not smell, acetylene is often described as smelling of garlic — it is an unpleasant, easily detected odour. Oxygen, on its own, does not burn, but it does greatly increase the rate of burning where ignition is already present. Acetylene is combustable and in a confined space it is explosive. A mixture of acetylene and oxygen in almost equal proportions burns at the welding torch nozzle with a flame temperature of around 3,000 degrees Centigrade (5,400 degrees Farenheit) and it should be noted that this temperature is fairly independent of the nozzle size being used by the welder.

Both oxygen and acetylene are stored in steel cylinders or bottles having a wall thickness of about ¼-inch — sturdy enough to withstand handling and transportation and the internal pressure (oxygen at 1980 - 2500 p.s.i.).

Welding is the process where separate pieces of the same metal are merged while molten to become one — steel and aluminium components can be welded, but steel cannot be welded to aluminium. Anyone other than a skilled welder will probably not progress beyond welding steel — the other metals which can be welded are more difficult and require slightly different techniques.

Brazing is not welding, since the metals are not merged to form one. It is a process where brass is melted between steel components and joins them when the brass cools in a process not unlike simple glueing.

THE DANGERS

You cannot take chances with gas welding and there are dangers which even the most safety conscious can only minimise. These dangers can be summarised as those connected with fire including burns to the operator and bystanders; eye damage through failure to wear effective goggles to protect the eyes from very harmful intensely bright light, sparks of red hot steel, and the risk of explosion when working with combustible gases in conditions where things can, and sometimes do, go wrong.

SAFETY PRECAUTIONS

Adequate welding goggles are an essential part of welding equipment. They should be used during any welding or brazing operation or irrepairable damage could be done to the eyes. Larger goggles for those who wear spectacles are readily available. When welding commences the darkened protective lenses must be brought down to protect the eyes. Welding goggles must be looked after and the lenses kept clean. Pitted or scratched lenses should be replaced.

Obviously there is a risk of fire or explosion with gas welding equipment. Before going down to the gas supplier and acquiring or hiring welding equipment, consider the safety aspect very carefully. Sometimes, no matter how careful you are, things can go wrong. Having worked in the restoration trade myself and after talking to Health and Safety officers and Fire Prevention Officers the following general points emerge:-

You should have one or more fire extinguishers and be prepared to ensure they are regularly serviced before you contemplate bringing home welding equipment. The sort of fires that might occur during welding range from petrol and oil, through rubber and plastics to fabrics. Water may work well for putting out trim or carpets that catch light, but it is highly dangerous where electricity or fuel and lubricant oil fires are concerned (oil floats on water), dry powder extinguishers can put out most automotive fires but must be properly looked after if they are to be relied upon. Foam extinguishers seem to be the fire experts' first choice, while the trade prefer dry powder.

Do not contemplate welding in a wooden building and make sure that you do not use the torch near inflammable materials such as petrol, paraffin or any painting materials.

I make a point of staying in the workshop for about twenty minutes after I have finished with the torch to make sure, as far as possible, that nothing is smouldering. To sum up all the safety precautions I would say that welders learn to have a healthy respect for the dangers and reduce all risks to a minimum. You should put in plenty of practice before trying any

(Continued)

repairs to a car, but bear in mind one golden rule – it pays to remove the petrol tank if you are welding anywhere in the region.

WHAT TO WEAR

I thoroughly recommend that you do not weld while wearing nylon overalls — cotton ones last much better and when a spark does hit them there is less danger of the hot fabric sticking to your skin. Wearing a woollen jersey below your overalls is not a bad thing but you should certainly not weld with bare arms or an open necked shirt because of the danger of sparks reaching parts they should not.

Leather industrial boots are going to give you more protection than ordinary shoes and if you wear plimsoles while welding the chances are that a spark will have you dancing energetically before long. Some welders favour wearing leather gauntlets when welding to protect the hands and wrists.

BURNS

If something does go wrong and you suffer minor burns they can be dealt with by dousing the affected area with cold water and applying paraffin gauze to the burn and then bandaging. If the burn is at all serious, and particularly if the skin is broken, you should put out the torch, turn off the gas supply and seek medical attention. Do not allow dirt to enter the burn.

WHERE YOU CAN WELD

There are no real restrictions about welding on private property as long as by doing so you do not contravene lease or tenancy agreements and the local authority does not take the view that you are conducting an industrial or commercial process in a residential area.

Those considerations aside, the worst place you can start gas welding is in a timber constructed garage containing two or three cars and slap bang next to a central heating storage tank. Having talked to representatives of Health and Safety and Fire Prevention officers it is quite clear that they would rather individuals did not weld in their domestic garages, but, if they must, then those bodies would like to see the work done in a stone floored and walled garage with an asbestos, or similar, roof. The building should be isolated from others, with all burnable materials removed and no car, other than the one being worked on, in the vicinity.

It makes sense to inform the local fire station that you have gas cylinders and will be welding, because in the event of a fire they will have some warning that your garage does contain a potential bomb.

Make no mistake about it — if fire reaches a gas cylinder you should prepare for an explosion that will have the effect of a wartime bomb as fragments of the cylinder casing fly a very considerable distance. If you think something has gone seriously wrong, try to shut the cylinders off, get everyone within a hundred yards well clear and phone the emergency services. *We are not exaggerating the dangers.*

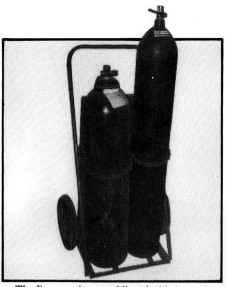

1 *The first step in assembling the kit is to place the cylinders vertically in a trolley or against a wall and secure them so that they cannot fall. Do not smoke when moving the cylinders or assembling the plant.*

2 *Remove any tape across the outlet of the cylinders and "Crack" them. This drastic-sounding process is nothing more than opening the delivery valve slightly, and for a very brief moment, to blow the threads clean before fitting the regulators.*

3 *The regulators are marked oxygen or acetylene; the first is marked blue and has a right-hand thread — the acetylene regulator marked red, has a left-hand thread.*

4 *Always use the correct spanner and never force the threads. No oil or grease muct come in contact with the threads.*

5 *I tighten all joints by hand before using the spanner. Here the blue oxygen hose is being attached to the regulator.*

6 *Tightening the joint between the hoses and the torch — note the provision of check valves to each hose to minimise the risk of flame travelling back to the bottles.*

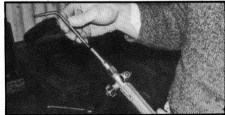

7 *Nozzles are numbered according to outlet size and I would stick to a 1 or 2 nozzle for practice purposes. The nozzle should be fitted hand tight, but not forced.*

EQUIPMENT

Most of the gas welding equipment available in the British Isles is supplied by BOC Ltd., commonly known as British Oxygen. The company have sales offices and supply depots throughout Britain and you can find the address of the nearest branch in your local telephone directory. BOC representatives will be pleased to discuss gas welding equipment and training courses with you, and can also provide technical assistance. You may also be able to obtain expert tuition from evening class courses in your area and I would thoroughly recommend a beginner to enroll because the instructors will be able to see what you are doing and tell you where you are going

wrong. They also know a good weld when they see one!

BOC will be able to sell you welding materials and equipment including goggles, gloves, hoses, gauges, torch and nozzles — they can also arrange supplies of gas under a hiring agreement which imposes various conditions. In general you will be unable to buy gas cylinders and must rent them. When you require re-fills some branches appear to allow you to take your cylinders to their depot and collect filled replacements, but at others (particularly depots run by agents) you may have to rely upon their transport — which costs more.

Being fully aware that there was a market

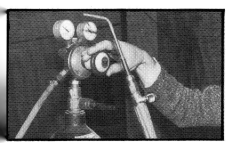

8 *Turn on the supply from each cylinder in turn and adjust the regulators with the torch tap opened. On this type of dual gauge regulator one dial indicates cylinder pressure and the other the delivery pressure to the torch. For normal purposes the beginner should equalise delivery pressures at or below 3 p.s.i. for oxygen and acetylene — this is done using the wheel at the front of the regulator.*

9 *After setting the delivery pressure close the torch taps and check all the joints are gas tight. Turn off one cylinder at a time and check the other line from the torch back to the cylinder using your senses of smell, touch and hearing — not a naked light. If you find a leak turn off the gas supply, dismantle and re-assemble the joint (do not use force) and if you cannot get a gas-tight joint, return the cylinder or offending part of the kit to the supplier.*

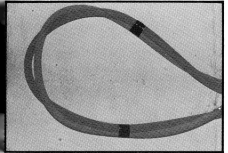

10 *Most welders strap the gas hoses together at about two feet intervals because they otherwise get entangled.*

11 *The torch can now be lit. Turn on the acetylene slightly and light it. Gradually increase the supply of acetylene which will initially burn with a watery, very sooty flame.*

12 *Increase the flame until it becomes ragged and the smoke stops (You do not need welding goggles on until you start welding).*

13 *Now gently turn on the oxygen supply and the flame becomes white and blue rather than orange-yellow.*

14 *As the oxygen is increased the outer edges become blue and a white cone forms at the tip of the nozzle. Now you have to strike a balance to achieve a neutral flame (equal parts of oxygen and acetylene being burnt). Insufficient oxygen will produce a long, ragged edged flame — too much will produce a small, blunt cone and a roar, and is known as an oxidising flame. Each type of flame has its applications, but you require a neutral flame for welding steel and that is what you should start working with.*

It may well be possible to hire gas welding equipment, but BOC do not encourage this practice. There is also a small black market in welding bottles and I would strongly recommend that you have nothing to do with it and on no account should you obtain gas cylinders filled by unofficial sources.

Welding equipment is not cheap although if used competently it can save you a lot of expense. I would prefer to buy new equipment and then look after it properly making regular inspections to ensure that each component is working safely and efficiently. If you do find damage to the hoses I would strongly advice you to replace them rather than attempt home repairs.

SETTING UP THE EQUIPMENT

I thoroughly recommend that before you go near welding plant you read and absorb the contents of one of the excellent instruction manuals issued by BOC and a second booklet entitled "Safe Under Pressure" — the company have taken a lot of trouble to provide easily understand guidance which is more or less "idiot-proof" (Do not go out and prove me wrong).

POPPING AND BACKFIRES

Popping can occur wnen the flame has an excess of acetylene — adjust the flame after checking that the nozzle end is clean and that there is not a blob of steel on the end. This shows up as a red glow when the torch is lit. Backfiring can often be the result of loose connections — more often than not a loose nozzle. Whatever the fault the torch should be turned off and the fault traced.

A substantial backfire is dangerous and may indicate that the backfire is taking place within the torch (it should not get hot in normal operation) which could allow flame to travel back up the hoses to the cylinders.

BOC strongly recommend that flame arresters are fitted at either end of hoses — I suggest that you follow their advice but do not rely blindly on the arresters. If something goes wrong shut off the cylinder valves and check the temperature of the acetylene cylinder.

SHUTTING OFF

In normal circumstances you should shut off the acetylene at the torch which will put out the flame and then turn off the oxygen. If the plant is to be left for more than ten minutes shut the cylinder valves and open the torch taps to release the gas in the hoses. If the plant is not going to be used for a longer period, unscrew the regulator control wheel until both gauges read nil. □

---

for low-cost, portable welding plants BOC introduced the portapack welding kit about three years' ago. This equipment employs small gas cylinders which are cheaper to fill and are rented to the customer on a long term low cost basis — in all other aspects the equipment is the same as and interchangeable with conventional welding equipment.

Below is a list of what basic equipment you will need and the approximate cost:

Gas cylinders	Annual rental:	oxygen	£22.08
		acetylene	£31.05
Regulators		oxygen	£29.44
		acetylene	£29.44
Spindle key			£ 1.27

Set of hoses (5 metres with check valves)	£15.76
Saffire DH welding torch	£52.21
Goggles	£ 5.06

Welding rods (sold by the packet, price varies according to thickness). The prices quoted were correct at the time of going to press and include VAT.

In addition to these basic essentials you may wish to purchase a pair of flashback arresters which will cost you £101.89, but a fire extinguisher is essential.

For the enthusiast/restorer the BOC Portapak is well worth considering since the price — £277.15 includes rental of the "mini" Cylinders for 12 years.

NEXT MONTH

We tell you some of the practice exercises and show how to start welding.

YOU TOO CAN GAS WELD

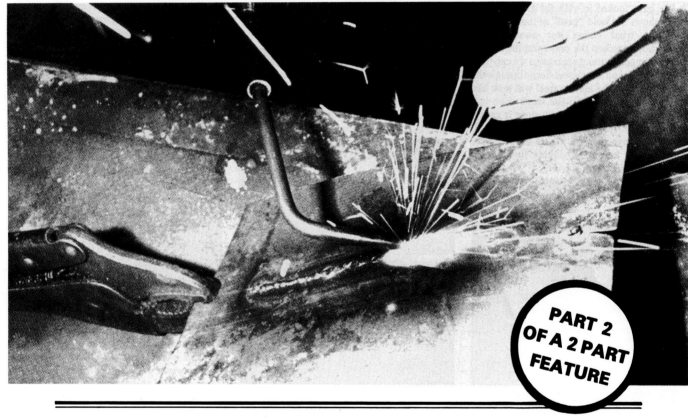

The exercises that will prepare you for your first welds and the problems that will arise when you start welding explained by Michael Brisby.

Last month we gave you the background information about gas welding plant. This month we describe the initial stages of learning to gas weld and describe some useful exercises.

Once you have got the hang of welding you will have to learn to use the torch while it and you are in all sorts of unlikely, and often uncomfortable, positions, but start by working at a bench. Protect the surface with a piece of ¼ inch steel plate if you can get it — never let the flame play on a stone surface because stone splinters will fly in all directions.

Lighting up.

To repeat what we said last month you should set the delivery pressures to about 2-3 lbs per square inch. With a number 1 nozzle or possibly a 2 fitted to the torch turn on the acetylene tap at the torch and light it before increasing the flow until the end of the flame becomes ragged and the smoke is visibly reduced. At this point gently turn on the oxygen tap and increase the flow — the flame becomes blue with a white centre and as the oxygen is increased a well defined white cone will form at the immediate end of the nozzle. A correct, neutral, flame is achieved with about equal parts of oxygen and acetylene being burnt and too much oxygen will tend to produce a blunt cone and a roar.

What a weld is

A weld is the fusion of like metals to form a joint between two components. This can be seen clearly when a good weld is sectioned — the consistency of the metal should be visibly constant throughout the joint, and this can only be achieved if the operator ensures that both items to be joined are molten at the point where the joint is required before any filling rod is introduced.

Molten Metal — The Pool

Take a piece of bright scrap steel — 18 gauge if you can get it — and place it on the bench. Now pull down your welding goggles and, holding the torch towards the back for the best balance and control, bring the flame to bear on the steel, lowering the torch until the white cone (which can be seen clearly through the goggles) is just above the metal surface.

Watch closely and you will be able to see the surface of the steel become molten after a

YOU TOO CAN GAS WELD

(Continued)

moment or two. Withdraw the torch and then try again.

We have looked in vain for a photograph which shows the weld "pool" of molten metal. We have tried taking our own and in desperation searched the photographic files of British Oxygen without success so we can only describe what you *should* see! Seen through the welding goggles the molten steel will look like mercury and if you keep the flame cone just off the surface of the steel and move it gently to one side (leftwards) the "pool" will obediently move ahead of the tip of the nozzle which should be angled slightly forwards in the desired direction of travel.

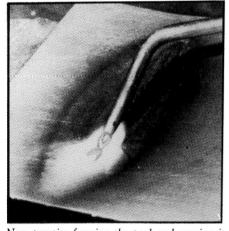

Now practice forming the pool and moving it across the surface of your piece of scrap steel from right to left. Most beginners find this exercise is much more difficult than it sounds and have to stop at regular intervals because of the concentration required. Keep at it because it is essential to be able to produce the pool, maintain it, and make it go where you want it to.

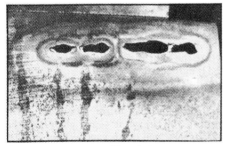

The most common problem is burning holes in the metal. This can be the result of using too large a nozzle in the torch, delivery pressures being too high or the flame being allowed to linger too long in one place. Keep practicing.

Examine each run across the steel — they should form a continuous concave grove on the surface of the metal.

Using Rod

When you have spent an hour or two practicing runs across a steel sheet and feel confident that you can produce the weld pool and guide it where you want, you can start using welding rod. The pictures show how.

Hold the rod in your left hand and form the weld pool before introducing the tip of the welding rod just above the pool. The rod will melt and merge with the pool without the rod contacting the sheet metal. The length of time that the rod is in the region of the flame and the weld pool will dictate the degree of weld build-up on the sheet steel (sometimes known as the weld bead).

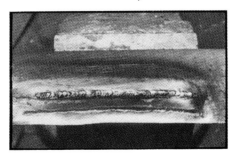

Here the concave track of a run without rod can be seen in the foreground and behind it the convex section of the run with rod added. A good run with rod will have a bright, blue-grey surface with a herringbone pattern.

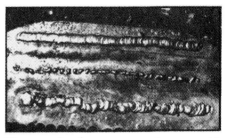

Three samples of attempts to make a run with rod. At the top is what a good run should look like — note the cross-section and surface pattern of the weld and its consistency. In the centre is a good example of a "non-weld" which is the result of applying molten rod without a weld pool on the parent metal. In the foreground the pool has been maintained inexpertly and it may be that the rod has been fed into the pool direct. There are also signs of the torch having wavered noticeably.

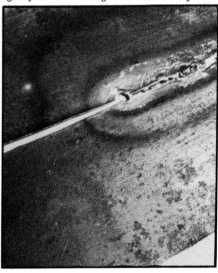

*When starting to weld pieces of metal together you will quickly find that parts being welded together **must** be clamped securely in position. These pieces of sheet metal were placed edge to edge and a weld started at one end without clamps or tack welds and the way the joint has dgstorted and spread can be seen.*

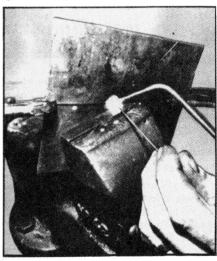

Using clamps and the vice the same joint is held in position while small welds (tacks) are made at intervals.

A different joint, but a clear indication of how the parts to be welded are tacked before commencing the weld proper.

Joining up the tacks.

Welding inside an angle is more difficult.

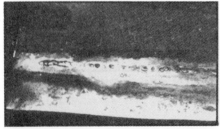

Allow the weld to cool before handling the metal or use pliers. Here the weld has been turned over to check proper weld penetration. At this stage you should test your practice welds to destruction the metal should tear before the weld parts.

A superb weld — what you should be aiming for.

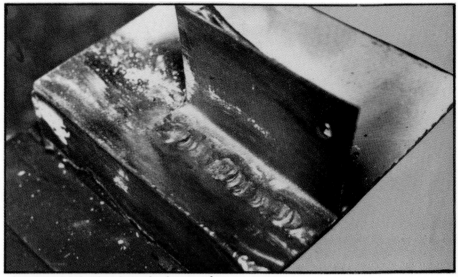

An example showing how the bottom sheet has tended to warp because of inadequate tacking and the unsightly, large weld that results.

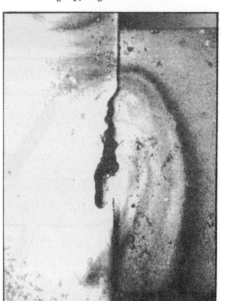

When welding one sheet laid on top of another like this the biggest problem is to get the lower sheet to heat up and the weld pool to form before the top component's edge melts away (known as under-cutting). This example gives a clear illustration of the problem.

An easy weld. Welding the outside of an angle like this is made easy because after making a few tacks the weld pool on each surface of the weld is readily obtained. With practice little or no rod will need to be added to form a strong weld on this type of joint.

(Continued)

When welding body panels the worst problem is distortion, and while you are still at the practicing stage try to achieve small, neat, welds and learn to counteract warping and distortion by tack welding, dressing the parts with a hammer to close up gaps, and using as little heat and welding rod as you can while producing a strong job.

Practice on pieces of scrap steel at the bench and then try the same exercises with the work clamped overhead and in other awkward positions you may meet when the time comes to work on a car. Do not attempt to weld anything on the car until you have tried a few repairs to a scrap panel or wing — there is a lot to learn and if you rush the process you will do more damage than good.

Before you try welding on a car get used to working in a variety of positions and attempt all sorts of joints. Repairing a scrap wing or panel is another useful exercise.

YOU TOO CAN GAS WELD

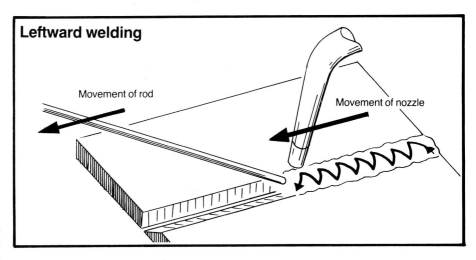

Leftward welding

Movement of rod

Movement of nozzle

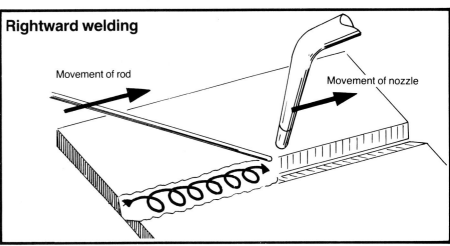

Rightward welding

Movement of rod

Movement of nozzle

There are two main methods of welding and the choice is determined by the thickness of the metal to be joined. Generally speaking, if the pieces to be welded are less than 5.00mm thick, as they are in most car work, the Leftward method should be adopted in which the weld proceeds from right to left along the joint. The torch should be kept moving at all times, but as well as the forward movement, there should be a slight zig-zag movement. This will ensure that both sides of the joint are kept suitably molten as the weld progresses. At the same time of course, the filler wire needs to be fed in to build up the body of the weld. The real trick is to keep your eye on the very centre of the weld pool and ensure that it is maintained. With practice it should be possible to hold the weld pool in a 'topped up' state as it moves along the joint, thus ensuring a satisfactory weld. The Rightward method of welding as you have probably already guessed, involves the weld moving from left to right. The filler rod is moved in circular movements along behind the torch and although this method is quicker and a little more economical, it is only applied to metal that is over 5.00mm thick, and so need not really concern us further.

The major problem that runs hand in hand with gas welding is heat distortion, and the thinner the subject, the greater are its effects. One precaution that can be taken to help reduce the effects of this distortion is to tack the joint prior to the final weld. The tacks should be made as neat and as small as possible because they can provide annoying obstructions to weld over if they are too large. On fine work with thin sheet metal the tacking is a very important part of the process, and should be set close together, but on thicker subjects the spacing can be greater.

If you can book yourself a place on a night school welding course do so, and for the very best training you could, if you can afford the fees, attend one of the British Oxygen Skill Centre courses at Waltham Cross, Manchester or Birmingham. (The BOC beginners' course lasts a fortnight and costs £200). If you can get yourself a place on a course make sure the instructor knows that you intend to weld cars and not girders or pipelines — if you cannot get on a course it is worthwhile finding a *good* welder who will tell you where you are going wrong. Good welders take a pride in their skills and I found they were often prepared to share their knowledge. Good luck! ☐

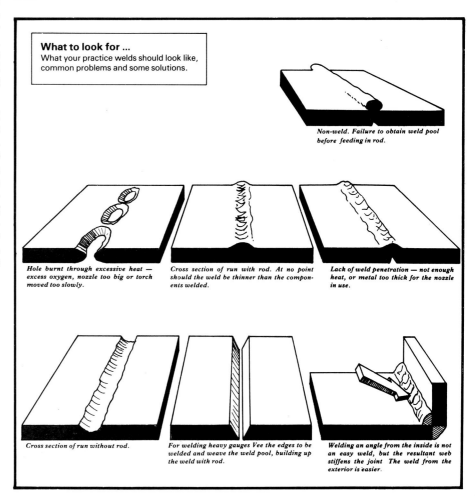

What to look for ...
What your practice welds should look like, common problems and some solutions.

Non-weld. Failure to obtain weld pool before feeding in rod.

Hole burnt through excessive heat — excess oxygen, nozzle too big or torch moved too slowly.

Cross section of run with rod. At no point should the weld be thinner than the components welded.

Lack of weld penetration — not enough heat, or metal too thick for the nozzle in use.

Cross section of run without rod.

For welding heavy gauges Vee the edges to be welded and weave the weld pool, building up the weld with rod.

Welding an angle from the inside is not an easy weld, but the resultant web stiffens the joint The weld from the exterior is easier.

stickyb

It is inevitable that a lot of the technological advances made specifically for the modern motor industry will eventually filter down and find applications within the classic car scene. Improved tyre technology, better oil and engine treatments and more effective paints, polishes and rust preventives are just some of the advantages now at our disposal. The clever thing is that, in a lot of cases, such technology can be applied to the classic car without outwardly effecting its originality. This gives the owner the best of both worlds. The car can be made safer, more efficient and longer lasting without ruining its classic appeal.

Panel bonding adhesives can provide a quick and very effective answer for sealing awkward joints.

One interesting new technique which I feel has great potential within our field is that of panel bonding with the tough, modern epoxy resin adhesives. It already has many applications within the modern motor manufacturing industry and the claims for its strength are impressive.

Down at the Mid-Kent College, Maidstone, Bryan Dunmall has been using panel adhesive for ages and is convinced of both its usefulness and convenience from a restoration point of view. Several years ago the college bought an expensive Wurth kit which he tells me is used to great effect on many of the classic projects which pass through the college's workshops.

This theoretical example simulates a repair made to a corroded door. The chalked area represents the rusty bottom edge which must be cut away and replaced with the section of new metal shown.

Is panel bonding an effective alternative to welding? – Chris Graham finds out

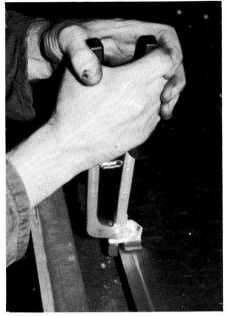

In this instance a joddled joint is used and it must be 15mm deep. The joddler is provided in the Wurth kit.

However, perhaps one of the greatest problems facing the future success of panel adhesive, in its battle with the conventional welded joint, is a psychological one. Professionals and DIY enthusiasts alike are bound to be sceptical about the relative strengths; on the face of it it is hard to imagine a 'glue' being able to fix a panel as strongly as a spot weld.

The arguments in favour of using a panel adhesive are certainly convincing. Such a product can be particularly effective for fixing repair sections on doors or wings for example. One of the risks when tackling this type of job with a spot welder is that careless work can lead to distortion in the panel. Using adhesive, of course, eradicates this risk completely because no intense heating is involved. It can also be a real boon if, for example, you need to fix an awkwardly-shaped inner wing to the outer wing and there is insufficient access for a spot welder.

Cleaning is a very important stage. An aerosol cleaner is provided in the Wurth kit and you must take time to use it properly. Any contamination left on the surface will reduce the efficiency of the bond.

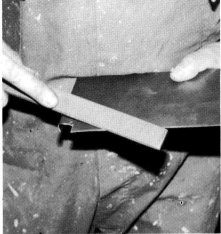

It's a good idea to file off the sharp edge on the repair section to make the final finishing stage easier.

In addition to providing a tough and resilient joint, panel adhesive also seals completely. A conventional spot-welded joint situated in the firing line of dirt and water thrown off a road wheel is bound, eventually, to leak. Paths will be forced through the joint in between the welds and, once the damp enters the supposedly protected area behind, corrosion will be inevitable.

While I am on the subject of corrosion there is another important point to be made here. Any form of metal welding, because of the extreme heat involved, alters the physical make-up of the metal. This leads to the setting up of a chemical imbalance which then initiates a gradual corrosive process. Such action will eventually weaken or even destroy the joint. The effects of this are usually checked with weld-through primers etc. but the risk is nevertheless still present. Applying the adhesive does not effect the metal or start any reaction at all and everything remains chemically stable.

With the repair section in place drill holes at about 3in intervals along the overlapping joint. Do not press too hard with the drill or you will distort the panel. Ensure the panel remains straight by fixing each hole as you go – self-tapping screws are quite adequate but Bryan used these fancy panel clips.

The adhesive is applied to cover the joddled joint. It only has to be applied to one surface.

Using panel adhesive is simplicity itself. The golden rules are that the panels must fit perfectly, a 15mm joddled joint should form the basis of the join when repair sections are being fitted, the bonding faces must be clean and free from grease and, once firmly fitted together, the joint must be left for about 12 hours – curing time can be reduced with heat lamps if required.

The tests we devised included a part-bonded and part-spot welded inner to outer wing joint, a textbook repair made to a simulated door skin and a straightforward test of strength between a bonded and a spot-welded joint. The picture series outlines the procedures involved in the practical testing. The strength test was carried out in the college's technical workshop but, unfortunately, I was not there to photograph it. However, the results proved to be very interesting. Two samples were prepared. Each joint was 100mm long and the spot welded example had a 12mm (standard prac-

Press the two panels together and secure through the drilled holes so that they are held tightly.

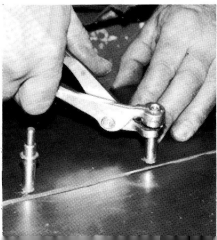

stickybusiness

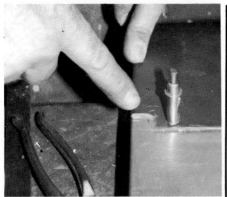

After about 12 hours (at normal room temperature) the adhesive will have set and a small MIG weld should be made at each end of the joint.

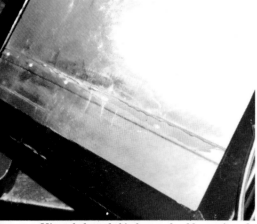

Viewed from behind you should see that the adhesive has been squashed through the joint to seal it completely – this is the desired effect. On the front of the panel excess adhesive should be removed before it sets.

tice) overlap secured by four evenly spaced welds. The bonded joint also measured 100mm but the overlap was slightly larger, at 15mm, as recommended by the manufacturers. As we expected, the testing machine was able to break both joints but we were interested to note that there was very little between the two of them. In fact, it was the bonded joint which proved to be slightly stronger, taking a force of 2.35 kilo Newtons before it broke. The welded joint snapped with 2.3 kilo Newtons showing on the gauge.

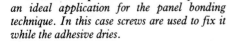

This wheelarch repair on a Ford 1600E provides an ideal application for the panel bonding technique. In this case screws are used to fix it while the adhesive dries.

Another example of a suitable candidate for panel bonding is this MGB GT. The joint between the rear wheelarch and the quarter panel can be stuck and sealed perfectly. The chalked areas on the two panels show where the bond will be made.

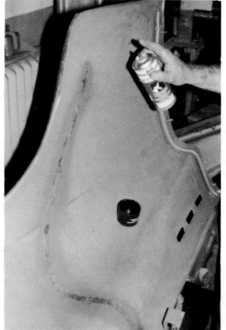

This is the inside of a front wing from a Sunbeam Talbot. The line being treated with cleaner marks the course of a joint between it and the inner wing.

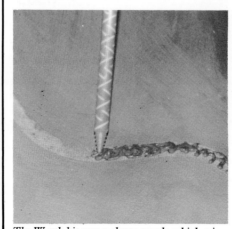

The Wurth kit uses a clever nozzle which mixes the two-part adhesive thoroughly before it reaches the tip.

The results speak for themselves and prove that panel bonding is a perfectly valid technique from a practical point of view. The one major drawback at the moment, as far as I can see, is the price. The Wurth kit owned by the college is currently selling for just under £400 and the tubes of adhesive cost £30 each. I have been in touch with Wurth UK Ltd who tell me that they do not market a kit aimed at the DIY market and that, in addition, they like to demonstrate the technique to the purchaser before making a sale.

While this kit is likely to be beyond the means of most of us working in the home workshop environment, it is perhaps worth bearing in mind the fact that bodies such as car clubs may wish to purchase a set. Most clubs are run on a sound economic footing these days and I know that extra services for the members are always worth considering. Owning this sort of equipment so that it can be hired out to members might provide a very useful and popular benefit.

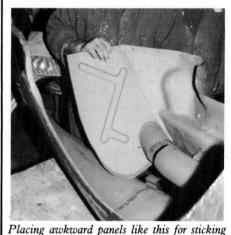

Placing awkward panels like this for sticking can be a two-man job. Correct alignment is obviously essential.

An alternative to drilling the joint is to use body clamps. These are preferable in this case.

I am sure that other manufacturers will shortly realise the potential of this technique and I know for a fact that one is already seriously engaged in production research. Once some competition is created and market forces drag down the price, panel bonding might become as commonplace as MIG welding as an aid to body repair. □

57

WELDING WITH

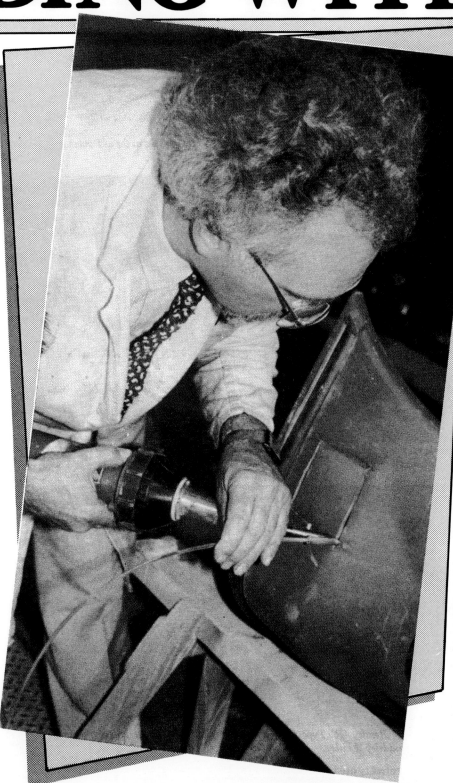

The idea of welding plastic together to make a repair is not quite as daft as it may first sound. The technique has been around for some years within the trade and now, with the rapid increase in the use of plastics for motor car construction, it is becoming ever more relevant. Interestingly, though, there are growing numbers of applications within the classic scene too. It is obvious to anyone who observes this market that, as time passes, progressively newer cars start to become collectable and with them come the modern manufacturing techniques. Dealing with this technology on a DIY basis requires a certain amount of new understanding which is where practices like plastic welding find their place.

When welding plastic you can seal up both splits or weld in patches in much the same way as you can with metal. If you are working with patches it is important that they be made from the same material as the item being repaired (you are likely to need a scrap item to take a sample from). The main reason for this is that disimilar plastics will have different melting points which will make the job harder and detrimentally effect the strength of the final weld.

This feature will illustrate exactly what is involved in the process and, with the help of Bryan Dunmall, I shall take you through a complete repair made on a plastic bumper. The techniques covered are equally suitable for use on cracked dashboards and other plastic panels too.

Before getting started, though, it is important to understand exactly what is involved. Basically, there are two groups of plastic to be considered. The first, thermoplastics, can be softened by heat and, therefore, are capable of being fused together in the form of a weld. In production these materials are heated and then press-formed into shape. If adjustments are required later then it can be re-heated and altered accordingly.

Thermo-setting plastics, on the other hand, undergo a non-reversible chemical change initiated by heat and pressure during production. This results in the formation of a hard, infusible material. The application of more heat will simply destroy rather than soften it, thus making it completely unsuitable for welding. Such plastics can be repaired only by bonding which is another story and something we will perhaps deal with in the future. Examples of thermo-setting plastics are Bakelite and GRP polyester resins.

Obviously it is important to be able to distinguish between the two types and one simple way of doing this is as follows. Cut a thin strip off your sample with a knife and note its reaction as you do so. A thermo-plastic will

Chris Graham and Bryan Dunmall assess the usefulness of plastic welding.

curl back as you cut, whereas a thermo-setting plastic will be removed as a straight strip.

Having established that your sample is indeed a thermoplastic and suitable for weld-

ing, the next thing is to find out exactly what type it is. There are many varieties and they all have different properties and, in particular, melting points. The plastic welding process involves the use of a filler rod and this

A DIFFERENCE

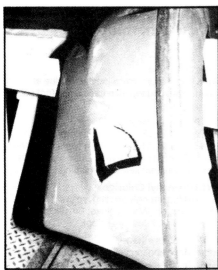

The first job is to assess the damage. Check for any distortion in the surrounding area. Any that is present may be corrected with gentle attention from the heat gun – don't overheat it. Bryan says that most plastic components have a 'memory' and that, given the chance, they will return to their original shape.

It is important to clean up the area to be repaired. It's much easier to fit a square repair section than one with jagged edges.

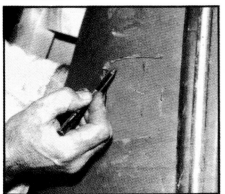

Hold the repair material (same type of plastic) behind the hole and draw round the edge to mark the size needed. The fit must be quite accurate.

Tin snips are quite adequate for cutting round the patch.

You will probably need to clean up the edge of the hole as well. Really, you need it as smooth as possible for the best results.

Several trial fittings will be required before you get a satisfactory fit. Small adjustments should be made with the file.

Next the joints have to be V'd out using an angle grinder so that room is created for the filler rod at the welding stage. However, it is important that only two thirds of the material thickness be removed so that the joint can be tacked into place while the alignment is checked etc.

This is the tacking nozzle being slotted on to the heat gun. The nozzle is heated by the air from the gun and is designed to be drawn slowly along the base of the V'd joint to fuse the two edges together.

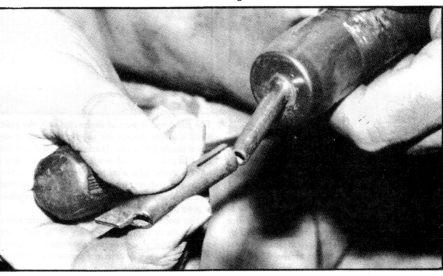

WELDING WITH

Once you have matched the welding subject with the appropriate filler rod you can establish the correct setting for the heat gun from the tables provided. This must be done at both the tacking and welding stages. The College's gun is made by Leister and costs about £210 + VAT. Sealey Power Welders Ltd produce one for £59.95 + VAT. (Tel: 0284 69621/701261)

Identification chart for commonly used thermoplastics by using the combustion method

(ABS) Acrylonitrile butadiene styrene:
Burns with a sooting, orange/yellow flame. Smells of rubber (not self-extinguishing).

(PA) Polyamide:
Burns with a sooting, yellowish flame, blue at bottom. Melts and produces foaming drops. Acrid smell, reminiscent of formic acid (self-extinguishing).

(PC) Polycarbonate:
Burns with a slightly sooting, yellowish flame, glowing ash. Has a sweetish smell (partially self-extinguishing).

(PE) Polythylene:
Burns with a clear, non-sooting flame, blue at bottom, yellow at top. Drips and smells like candle grease (not self-extinguishing).

(PMMA) Acrylic:
Burns with a clear, crackling flame, blue at bottom, yellow at top. Aromatic smell (not self-extinguishing).

(PP) Polypropylene:
Burns with a clear, non-sooting flame, blue at bottom, yellow at top. Squirts a wave of molten material in front of the flame. Drips and smells like oil or wax (not self-extinguishing).

(PVC) Polyvinyl Chloride:
Burns with a strongly sooting, yellow flame, green at edges. White smoke. Smells of hydrochloric acid (self-extinguishing).

(TPUR) Polyurethane:
Burns with a sooting yellow flame, blue at the base. Has a sweetish smell, drips like candle grease (partially self-extinguishing).

Table reproduced by courtesy of the Motor Insurance Repair Research Centre

Bryan recommends that you tack two opposite sides at a time. The nozzle should always be pulled and never pushed. Pressing down too hard with the nozzle will distort the joint, so be careful. Faults in alignment can be corrected with more gentle heat from the gun

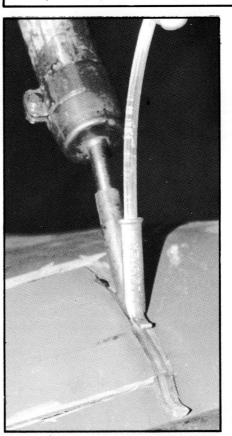

Once the heat level has been set you can fit the appropriate welding ('Speed') nozzle. The one to fit will be determined by the section of the filler rod being used (it can be triangular, round or flat). The gun and nozzle must be allowed at least three or four minutes to warm thoroughly before you start.

The filler rod is passed down through the guide until it meets the hot air supply where it is softened ready for fusion. It continues on down into the V'd groove which is also being heated and a joint is made. Bryan advises that you start laying down the rod just ahead of the groove so that, when you actually drop down into it, everything is good and hot and working properly.

You must keep a gentle pressure on the rod so that it feeds through smoothly as it melts and draw the gun along keeping the nozzle just above the surface. Any downward pressure will distort it.

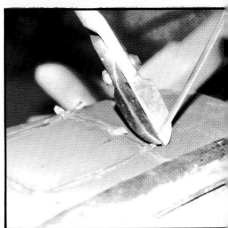

When you reach the end of a run lift the gun up and off the remainder of the filler rod and snip the excess rod off neatly. Single cracks are treated in the same way, although you may need to weld them from both sides to give the necessary strength.

With all four sides welded the patch should look something like this. If you find difficulty in getting the weld to fuse then there may well be oxidation problems. To overcome this, give the area a good clean and start again.

A DIFFERENCE

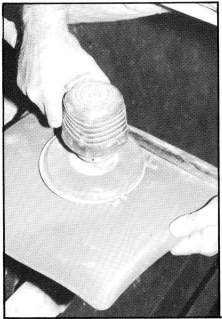

Final finishing involves sanding by hand or with an orbital sander like this. Any imperfections, in fact, can be filled with a suitable flexible plastic filler (3M produce such a product).

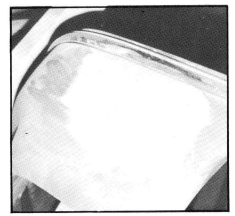

The finished repair. If you have been repairing a bumper with a textured finish this will have been removed by the sanding and will have to be replaced with the proper textured finish paint (for instance Formula 40 Bumpercoat Textured). Dashboards will need to be completely repainted to hide the repair.

The important thing to remember regarding painting is that neither cellulose or synthetic paints can be used – it must be two-pack or Formula 40 Bumpercoat. Bryan recommends that you get the item professionally sprayed once you have finished the weld and prepared the surface.

must be made from the same material as that being welded if the best results are to be obtained. Therefore, to select the correct rod you need to know exactly what you are welding. Bryan tells me that there are several directories which list cars and the types of plastics used on them. However, these usually relate just to the modern car market which limits their usefulness from our point of view.

Fortunately, though, there are a couple of more practical alternatives. The first is a solvent test using a special kit. Answers are arrived at by a process of elimination using the various solvents included. Bryan considers that such kits are quite expensive (about £20) but, if you want one, he thinks that Berger produce them. The other method is a simple flame test. You need a gas lighter and a small strip of the plastic under investigation. Allow the flame to play on the strip and watch for the characteristic signs as it burns (see table). This is fairly accurate although it should be noted that some of the reactions are rather similar and this can lead to slight confusion.

If, for whatever reason, you cannot locate the correct rod it is possible to use a length cut from your scrap piece to use as a substitute. However, the efficiency of this compromise can sometimes be effected by the fact that most plastics are not completely pure. The advantage of the genuine filler rod is that it is totally pure and so produces the best possible joint. □

A cut above the rest

To the many of us who have struggled for hours with tin snips, hack saws, hammers and chisels, grinders and even air saws, a plasma cutter represents heaven on earth. It is quick, simple, accurate and very effective. Sills can be removed in literally seconds and the cut produced is neat and, more importantly, distortion-free.

However, not until recently have the prices of these ingenious tools started to drop into range of the DIY restorer. Manufacturers are now finding themselves able to offer their smaller models for sale at about £400, which should open up a whole new market. At last the small business and even the keen enthusiast can revel in the luxury afforded by a plasma cutter. It will bring to them the possibility of saving many hours of hard physical work with the added bonus of producing a better and more efficient result into the bargain.

Plasma cutting is a technique which we have never featured before in *Practical Classics & Car Restorer* and I would imagine that quite a number of you are not entirely sure how it works. Most people probably associate plasma with blood and hospitals but the word has another meaning as well. It also refers to a gas which has become ionised (negative electrons are split from the atom which then becomes positively charged) due to extreme heat. Because of this ionisation the gas (ordinary air in this case) becomes able to conduct electricity and it is this phenomenon which makes the whole process possible. In simple terms plasma cutting occurs when a stream of plasma is used to convey an electrical arc, in a fine jet, down on to the surface being cut and the metal is literally blown away.

To operate most plasma cutters successfully it appears that you will need a heavy duty mains supply providing at least 20A at 240V. A 32A single-phase switched socket outlet (which costs about £50) will have to be fitted and to do this you must run some very heavy duty cable from the mains distribution board. Having this professionally installed can be quite expensive.

Another requirement is a decent compressor to supply the large volume of air (ideally 6cfm at 70psi) for the plasma stream and this is attached to the unit via an airline and regulator valve/filter assembly on the back. Compressed air is fed down the cable to the handset and straight out through tiny holes in the nozzle. The heat required to ionise the air comes from an electric arc which jumps between two electrodes in the nozzle at the moment when the trigger is pressed. Instantaneously the air becomes charged and is then able to support the arc. To complete the circuit and encourage the arc's passage down the plasma stream and on to the work an earth clamp is added. This is located on a conveniently clean area in exactly the same way as you would do when MIG welding.

The stream of plasma 'containing' the arc passes down through the small hole in the tip to form a jet and this intensifies its effect. The temperature at the business end of a plasma cutter is more than 24,000 degrees C – quite sufficient for most jobs! This intense,

When using a plasma cutter gloves, goggles and clean cotton overalls are essential. For these tests we used the SIP Plasma 25 (priced at about £400) which is as portable as any small MIG welder.

Chris Graham looks at the technique of plasma cutting.

localised heat melts the metal and the pressure of the plasma jet blows it away in a dramatic shower of sparks.

Using a plasma cutter

To help with the practical side of this feature Ian Callander of SIP arranged for one of their machines to be delivered to Bryan Dunmall at the Mid-Kent College in Maidstone, Kent. Bryan and I then set about experimenting with it to judge for ourselves the practicalities of this technique. The following account obviously relates specifically to the SIP Plasma 25 but we imagine that most of it will be generally true of the other machines currently on the market.

One of the beauties of using a plasma cutter like this SIP unit is that there really is very

A cut above the rest

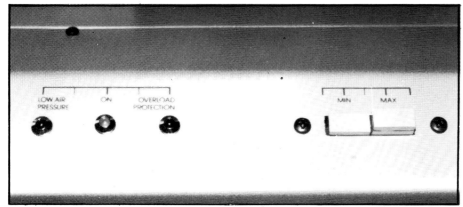

The controls on the SIP Plasma 25 are very simple. The three lights on the left indicate low air pressure, overload protection and that the machine is switched on and ready to use. There are only two controls for the power, min or max (on the right). Further to the right, out of shot, is the main on/off rocker switch.

Incidentally, with the air flowing, take the trouble to look at the dial on the valve assembly and check that the correct pressure has been set. Select the power setting required for the thickness of metal being cut and place the torch at the point where the cut is to start. It is important that the tip makes good contact with bare metal at the start because, if not, the arc will fail to strike.

By this stage the air should have turned itself off again (usually after about 30 seconds) and, actually to start cutting, the trigger must be pressed twice in quick succession. The first push will turn on the air again and the second, which should be no more than one second after the first, will activate the electrics to strike the arc. This triggering action has been designed for two important reasons and the first is safety. The idea is that

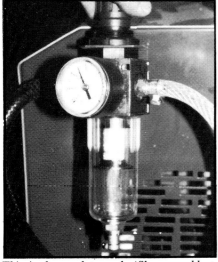

This is the regulator valve/filter assembly to which the air supply from the compressor is attached. Adjustment to the pressure setting is made with the knob at the top; the glass bowl below collects water.

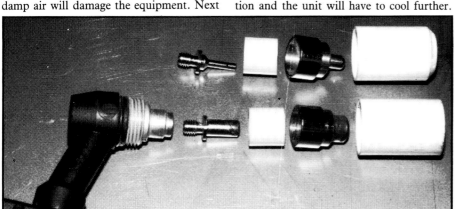

Plasma cutting provides the ideal way of making a perfectly accurate repair section. Select the damaged area and clamp a section of correctly shaped new sheet over the top. Mark on the top section the shape to be cut...

little to do before you actually start work. The first task is to connect up the air supply from your compressor. At this stage it is also worth checking the collection bowl on the regulator valve/filter assembly for any water. If found it should be drained away because damp air will damage the equipment. Next connect the earth clamp to the work ensuring that an effective contact is made and then switch on at the mains.

As an initial test, press the trigger once and the air flow should start. If this does not happen either there is insufficient pressure (adjust the valve) or, due to recent and excessive previous use, the thermal cut-out is in operation and the unit will have to cool further.

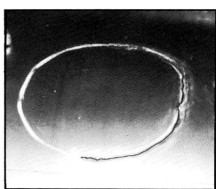

...and run round the line. As long as you have made the patch over-sized then slight 'wobbles' in the cut do not matter. In fact they act as registration marks when positioning the new section.

the double action will guard against accidental operation of the unit. The second is cooling. The air is switched on by the first press so that it is made impossible for the electrodes to become overheated and consequently damaged. Taking your finger off the button cuts the electrical supply but the air flow continues afterwards to ensure adequate cooling.

The nozzle of this plasma cutter consists of (from left to right) central electrode, ceramic insulator, external electrode and ceramic heat-resisting nozzle cover. The set above features narrower electrodes for working in tight corners.

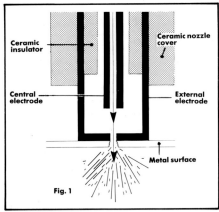

Cross-section through nozzle showing the passage of air stream down through hollow central electrode and out through hole in external electrode.

It took Bryan about 25 seconds to cut this out. The section below represents the old panel which can be thrown away, leaving the new piece to drop exactly into place. The cutting process leaves a gap of about 1mm which is ideal for butt welding afterwards.

Immediately you start a hole will be made in the metal and movement along the desired path can begin. Failure to cut completely through the surface means that you are travelling too fast or that the metal is too thick for the machine to cope with.

Bryan concluded that this machine cut at a rate some two or three times faster than any other conventional sheet metal cutting device which he had tried. This, he said, was typical of all plasma cutters and he was most impressed. However, he was quick to point out that practice was, as usual, essential. Because the handset is held vertically when in use the arc is completely obscured from view. This can make it tricky for the beginner accurately to follow a line. One simple solution, of course, is to use a template and this would provide a virtually failsafe way of cutting an accurate path.

Applications and safety

Perhaps the greatest single use for the plasma cutter is for the removal of car body panels. However, great care is needed when working on double-skinned areas. The SIP Plasma 25 was capable of bridging gaps of 10mm between panels while still remaining completely effective. Therefore, it would not be suitable for removing a corroded outer wheelarch section if the inner panel was to be left in place. Bryan considers that all the other similarly priced units will show these same characteristics. To overcome this limitation he added that some of the most expensive machines feature a plasma arc length control which enables the user to set the 'depth' of the cut and thus avoid damaging anything other than the surface being treated.

Problems can arise when encountering areas of thickly layered filler. It will quite happily cut through lightly filled surfaces but, once the layer becomes more than about 4mm thick, the machine is likely to cut out. Bryan thinks that one reason behind this might be that filler is a thermo-setting plastic which doesn't melt but gradually decomposes with heat. This slows down the progress of the plasma cutter to such an extent

that the circuit is broken which brings everything to a halt. The simple remedy, of course, is to run along the filled area with a grinder to produce a line of bare metal along which to work.

The plasma cutter scores with its high cutting speed and ease of manipulation. Only a lightly tinted pair of goggles is required

The plasma cutter works just as well on uneven surfaces. The cutting range for this set is anything between 2 and 5mm. We tested it up to 4mm and it worked beautifully.

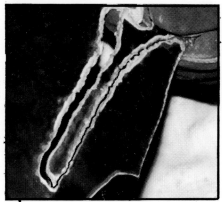

Painted surfaces do not stop a plasma cutter either; it is all blown away. If cutting a painted panel you will need to start on bare metal to get the arc going.

(similar to oxy-acetylene goggles) which means that you can follow a marked line without difficulty.

Fire is always a risk with this type of work although with a plasma cutter it is minimal. As long as you are sensible and avoid cutting through petrol tanks, fuel lines, brake pipes etc. then the risk of a fire will be slight. In most cases the metal should not stay hot for long enough to establish a fire. Painted and undersealed surfaces can be cut without problems; nevertheless, Bryan advises that all trim be removed from the immediate area and, of course, only a fool would leave the fuel tank closer than a stone's throw from a plasma cutter. Generally speaking you should apply the same anti-fire rules to plasma cutting as you do to MIG welding.

Bryan also considers it wise to operate a plasma cutter only in a well ventilated area although he added that the fumes produced were not so nasty as those given off when cutting with oxy-acetylene equipment. However, particular metals react differently to the process and you should look out for zinc- or cadmium-coated metals because toxic gases are produced by these. Also be wary of the presence of any solvent or de-greasing agents on the surface being cut because these too can lead to the production of dangerous gases.

There is no question that a plasma cutter would be a welcome addition to any small workshop which undertakes a lot of body repairs. However, for the DIY enthusiast the expensive requirements might, in some cases, outweigh the advantages of such a tool. I can imagine owners finding it hard to justify the expense for the restoration of just one car. Nevertheless, clubs might see the possibilities of purchasing one and providing another useful service to their members. Perhaps sometime soon a manufacture will launch a unit to run off a 13A supply which will put a different complexion on the whole picture. Until then though it is likely that most will regard the plasma cutter as an expensive but very desirable luxury. □

Welding Equipment Mail Order Availability Chart

JANUARY 1992

Supplier (Tel)	BOC GAS WELDING EQUIPMENT	CEBORA WELDING EQUIPMENT	CLARKE WELDING EQUIPMENT	CLARKE PLASMA CUTTERS	KEL-ARC WELDING EQUIPMENT	MAYPOLE WELDING EQUIPMENT	MIGATRONIC WELDING EQUIPMENT	MOTIVAIR WELDING EQUIPMENT	SIP WELDING EQUIPMENT	SIP PLASMA CUTTERS	SUREWELD 'MONOMIG'	TELWIN WELDING EQUIPMENT
A B TOOLS — Tel: 0782 566222		●										
CAMBRAI MARKETING — Tel: 0269 595209						●	●		●	●		
EASTWOOD COMPANY — Tel: FREEPHONE 0800 897303					●							
FLAIRLINE LIMITED — Tel: 081-450 4844					●				●	●		
FROST AUTO RESTORATION — Tel: 0706 58619					●							
ROY HOWE WELDING EQUIPMENT — Tel: 0325 356900	●	●				●			●	●		
T F KELLER & SON — Tel: 0603 663790					●							
LANCING WELDING SUPPLIES — Tel: 0903 763152		●							●	●		●
MACHINE MART — Tel: 0602 587666			●	●								
MOTIVAIRE LIMITED — Tel: 0992 714471								●				
PHIL WEEKS WELDING — Tel: 0225 312177	●						●		●	●		
POWER TOOL CENTRE — Tel: 051-424 4545									●	●		
PREMIER WELDING CENTRE — Tel: 061-872 8579	●											●
SUREWELD — Tel: 0933 57005											●	
TOOLWISE LIMITED — Tel: 0533 681175									●	●		
TRANSPEED LIMITED — Tel: 0273 774578		●							●	●		
WELDING CENTRE — Tel: 081-399 2449	●					●	●		●	●		
WELDMET LIMITED — Tel: 081-947 1288	●											
WELDING WORLD — Tel: 0282 871517	●		●	●					●	●		

TRANSPEED

213 PORTLAND ROAD, HOVE, SUSSEX BN3 5LA
SALES HOTLINE 0273 774578

We have one particular customer (who is a qualified Coded welder in all major metals) who has purchased much welding equipment from *TRANSPEED* over the years. We were particularly baffled, however, by his rather excessive purchasing habits on the **Cebora 225 Pocket Tig**; namely, three machines in just under six months! Somewhat anxious that reliability or durability might be in question, we asked him if his previous machines were OK. "Yeh – fine, mate! My pals at the workshop liked the small unit's performance so much they bought 'em off me for mobile use." Phew! What a relief. I think we can take that as a recommendation...

Cebora 225 Tig

Cebora 883 Portable Mig

The two **Cebora Mig** models (883/886 110amp/130amp respectively) have been very strong sellers for *TRANSPEED* for over six years. Interestingly, during that entire period, we can recall only one small modification to the wire speed control circuit. The rest of the machine has remained the same throughout. Some manufacturers might regard this as a lack of Research & Development but, to our view, why change something that is right in the first place? As soon as you use one of these machines you will realise you have a thoroughbred. Quality performance, easy to use and maintain.

This machine is the hallmark of portable air compressors. *TRANSPEED* have featured this **TRANSAIR** machine for over five years. One of these machines has been giving sterling service to our Motorsport Team, regularly taken around the UK from circuit to circuit providing compressed air in the Pits for wheel wrenches, tyres, drills, etc. With over 450 units of this model sold so far, we at *TRANSPEED* know that for 4.5cfm of Free Air Delivered with quality durability, the price is right!

Transair Compressor

Look! **£149.**95 + VAT

Now if you want a really *serious* duty cycle for welding yards and yards of chassis rail or heavy gauge materials, but with sufficient control to weld thin body sills too, the **SIP Autoplus 130** is the machine that will do it!

FEATURES: ● EURO CONNECTOR TORCH ● 6 POSITION POWER CONTROL WITH AUTO WIRE SPEED LINKED TO POWER CONTROL ● FAN COOLED FOR EXTRA DUTY CYCLE ● SOLID STATE ELECTRONIC CONTROL SYSTEM AIDS QUALITY WELDS.

SIP Autoplus 130

TRANSPEED CONSUMABLES SERVICE

It is also *TRANSPEED* policy to ensure that customers who purchase our machines are not left high and dry when it comes to spares and consumables. There are spare Mig tips from 85p, shrouds from £4.25, umbilical liners from £4.95 , gas regulators from £17.50. For your added safety we strongly recommend the leather gauntlets (£3.95pr) and Welding Masks from £8.95. We even stock a range of competitively priced Fire Extinguishers. For your convenience, we have a complete range of consumables such as Abrasive Discs, Arc Rods from £1 per 25, Steel Mig Wire from £3.95 and Gases (collection only) from £6.95.

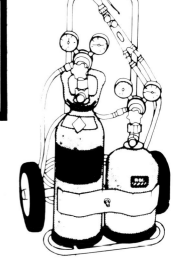

67

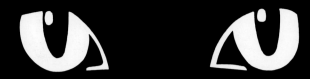

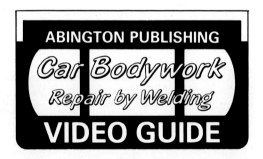

68

SPONSORS OF THE PRACTICAL CLASSICS TR6 REBUILD PROJECT

69